Automated Verification of Concurrent Search Structures

Synthesis Lectures on Computer Science

Science Fiction Prototyping: Designing the Future with Science Fiction
Brian David Johnson
2011

Storing Clocked Programs Inside DNA: A Simplifying Framework for Nanocomputing
Jessica P. Chang and Dennis E. Shasha
2011

Analytical Performance Modeling for Computer Systems
Y.C. Tay
2010

The Theory of Timed I/O Automata
Dilsun K. Kaynar, Nancy Lynch, Roberto Segala, and Frits Vaandrager
2006

Automated Verification of Concurrent Search Structures

Siddharth Krishna, Nisarg Patel, Dennis Shasha, and Thomas Wies

ISBN: 978-3-031-00678-4   paperback
ISBN: 978-3-031-01806-0   ebook
ISBN: 978-3-031-00074-4   hardcover

DOI: 10.1007/978-3-031-01806-0

A Publication in the Springer series
SYNTHESIS LECTURES ON COMPUTER SCIENCE

Lecture #13
Series ISSN
Synthesis Lectures on Computer Science
Print 1932-1228   Electronic 1932-1686

Automated Verification of Concurrent Search Structures

Siddharth Krishna
Microsoft Research, Cambridge

Nisarg Patel
New York University

Dennis Shasha
New York University

Thomas Wies
New York University

SYNTHESIS LECTURES ON COMPUTER SCIENCE #13

ABSTRACT

Search structures support the fundamental data storage primitives on key-value pairs: insert a pair, delete by key, search by key, and update the value associated with a key. Concurrent search structures are parallel algorithms to speed access to search structures on multicore and distributed servers. These sophisticated algorithms perform fine-grained synchronization between threads, making them notoriously difficult to design correctly. Indeed, bugs have been found both in actual implementations and in the designs proposed by experts in peer-reviewed publications. The rapid development and deployment of these concurrent algorithms has resulted in a rift between the algorithms that can be verified by the state-of-the-art techniques and those being developed and used today. The goal of this book is to show how to bridge this gap in order to bring the certified safety of formal verification to high-performance concurrent search structures. Similar techniques and frameworks can be applied to concurrent graph and network algorithms beyond search structures.

KEYWORDS

verification, separation logic, concurrency, data structures, search structures, B trees, hash structures, log-structured merge trees

Contents

Acknowledgments

We would like to acknowledge the very careful external reviews by Maurice Herlihy, Eddie Kohler, Robbert Krebbers, K. Rustan M. Leino, and Peter Müller. In addition, we warmly acknowledge Elizabeth Dietrich and Rafael Sofaer for their careful reading during the preparation of this manuscript. We are also thankful to Alexander Summers for his contributions to the development of the flow framework, which is one of the pillars of our verification technique. Because our work was built largely on Iris, we would like to acknowledge the support received from the Iris Helpdesk, especially from Ralf Jung and Robbert Krebbers. Much of this work has been funded by the National Science Foundation under grants CCF-1618059, CCF-1815633, MCB-0929339, and NYU Wireless. The first author is grateful for the support of Microsoft Research Cambridge and his research team. Our publisher, Diane Cerra of Morgan & Claypool, has been very understanding with our obsessive natures. In addition to Diane, we would like to thank Christine Kiilerich, C.L. Tondo, and Brent and Sue Beckley.

Siddharth Krishna, Nisarg Patel, Dennis Shasha, and Thomas Wies
April 2021

CHAPTER 1

Introduction

For the last 60 years or so, the processing power of computers has been doubling approximately every two years. For most of that time, this growth has been backed by the increase in the number of transistors present on integrated circuit chips, a phenomenon commonly known as Moore's Law. Until about 2000, this was accompanied by a corresponding increase in clock speed, the speed at which computers perform each step of computation. However, in the early 2000s, computer hardware started reaching the physical limits of clock speed, mostly due to heating and quantum effects. To counter this, manufacturers have turned to parallel architectures where increased transistor densities are being used to provide multiple cores on a single chip, enabling multiple computations to be performed in parallel.

Having n processing cores on a chip, however, does not immediately imply a factor of n increase in speed. To make the most of these multicore machines, software needs to be carefully designed to efficiently divide work into threads, sequences of instructions that can be executed in parallel.

Well-designed parallel algorithms distribute the workload among threads in a way that minimizes the amount of time each thread spends waiting for other threads. A standard way to achieve this is to store any shared data in so-called concurrent data structures, supported by multi-threaded algorithms that store and organize data. These data structures are now core components of critical applications such as drive-by-wire controllers in cars, database algorithms managing financial, healthcare, and government data, and the software-defined-networks of internet service providers. The research and practitioner communities have developed concurrent data structure algorithms that are fast, scalable, and adaptable to changing workloads.

Unfortunately, these algorithms are also among the most difficult software artifacts to develop correctly. Despite being designed and implemented by experts, the sheer complexity and subtlety of the ways in which different threads can interact with one another means that even these experts sometimes fail to anticipate subtle bugs. These bugs can cause the data to be corrupted or the program to misbehave in unexpected ways. For instance, consider the standard textbook on concurrent algorithms, *The Art of Multiprocessor Programming* [Herlihy and Shavit, 2008]. Although written by renowned experts who have developed many of the most widely used concurrent data structures, the errata of the book list several subtle but severe errors in the algorithms included in the book's first edition. There have also been many such examples of concurrent algorithms in peer-reviewed articles with (pencil-and-paper) mathematical proofs that have later turned out to contain mistakes [Burckhardt et al., 2007, Michael and Scott,

1995]. Therefore, it is clear that we need systematic and dependable techniques to reason about and ensure correctness of these complex algorithms.

Formal verification aims to use mathematical techniques to prove, in a rigorous and machine-checkable manner, the absence of bugs and the conformity of a system to its intended specification. Several projects have demonstrated the successful use of formal verification to improve the reliability of real-world software designs. These success stories roughly fall into two groups. At one end are projects that use interactive or semi-automated theorem provers to verify that a software system meets its functional correctness specification. Prominent examples are the verified optimizing C compiler CompCert [Leroy, 2006], the verified OS microkernel seL4 [Klein et al., 2010], and the verified HTTPS replacement developed in Project Everest [Bhargavan et al., 2017]. The degree of reliability achieved by these systems is impressive. For instance, Yang et al. [2011] reported that out of 11 commercial and open-source C compilers, CompCert was the only one that did not exhibit wrong code generation errors when exposed to extensive random testing. However, the human effort needed to establish functional correctness proofs that provide such high reliability remains tremendous. For each of the above projects, it involved several years of work done by verification experts.

At the other end are automatic static analysis and model checking tools. Notable examples are the ASTRÉE analyzer [Blanchet et al., 2003] for verifying the absence of run-time errors of embedded systems code, which has been applied to large code bases in the avionics industry, and Microsoft's Static Driver Verifier [Ball et al., 2011] for verifying domain-specific properties of Windows device drivers. More recently, Amazon and Facebook have started to incorporate such tools into their continuous integration tool chains, automatically analyzing millions of lines of code written by ordinary software engineers [Chong et al., 2020, Distefano et al., 2019]. However, to achieve such a high degree of automation and scalability, static analysis tools target generic correctness properties (e.g., absence of run-time errors) rather than full functional correctness, or focus on detecting errors rather than proving their absence. Therefore, these tools can still miss intricate errors in the analyzed software.

Against this backdrop, this monograph aims to bring the certified safety of formal verification to *concurrent search structures*. A search structure is a key-based store that implements a mutable set of keys or, more generally, a mutable map of keys to values. It provides five basic operations: (i) create an empty structure, (ii) insert a key-value pair, (iii) search for a key and return its value, (iv) delete the entry associated with a key, and (v) update the value associated with a particular key.

We demonstrate a strategy that reduces the manual effort involved in establishing functional correctness proofs of the concurrent search structures that are in use every day. We achieve this reduction by increasing the modularity of the proof argument and by making the proofs reusable across diverse data structure implementations. While we do not aim for fully automated verification here, we hope that the presented techniques will also inform the design of future automated static analysis tools by providing a uniform framework for reasoning about the correctness of concurrent search structures.

1.1   ALGORITHMIC MODULARITY

Concurrent algorithms are complex because they have to perform two very difficult tasks simultaneously: manage interference among threads in such a way as to ensure correctness, and organize the data in memory so as to maximize performance. The resulting combination of delicate thread protocols and advanced data layouts used by concurrent search structures makes formally verifying them challenging.

Modularity is as important in simplifying formal proofs as it has been for the design and maintenance of large systems. There are four main types of modular proof techniques: (i) Hoare logic [Hoare, 1969] enables proofs to be compositional in terms of program structure; (ii) local reasoning techniques [Banerjee et al., 2013, Kassios, 2006, Müller, 2001, O'Hearn et al., 2001, Reynolds, 2002] allow proofs of programs to be decomposed in terms of the state they modify; (iii) thread modular techniques [Herlihy and Wing, 1990, Jones, 1983, Owicki and Gries, 1976] allow one to reason about each thread in isolation; and (iv) refinement techniques [Back, 1981, Dijkstra, 1968, Hoare, 1972] enable reasoning about properties of a system at different levels of abstraction.

Concurrent separation logics [Brookes, 2007, Brookes and O'Hearn, 2016, da Rocha Pinto et al., 2014, Dinsdale-Young et al., 2013, 2010, Dodds et al., 2016, Feng et al., 2007, Fu et al., 2010, Jung et al., 2015, Leino et al., 2009, Nanevski et al., 2014, O'Hearn, 2007, Svendsen and Birkedal, 2014, Vafeiadis and Parkinson, 2007] incorporating all of the above techniques have led to great progress in the verification of practical concurrent search structures, including recent milestones such as a formal paper-based proof of the B-link tree [da Rocha Pinto et al., 2011].

An important reason why many proofs, such as that of the B-link tree, are still complicated is that they argue simultaneously about thread safety (i.e., how threads synchronize) and memory safety (i.e., how data is laid out in the heap). We contend that safety proofs should instead be decomposed so as to reason about these two aspects independently. When verifying thread safety, we should abstract from the concrete heap structure used to represent the data and when verifying memory safety, we should abstract from the concrete thread synchronization algorithm. Such an approach promises reusable proofs and simpler correctness arguments, which in turn aids proof automation.

As an example, consider the B-link tree, which uses a forward pointer-based technique for thread redirection. The following analogy [Shasha and Goodman, 1988] captures the essence of the link-based technique on a macroscopic data structure.

Bob wants to borrow book k from the library. He looks at the library's catalog to locate k and makes his way to the appropriate shelf n. Before arriving at n, Bob gets caught up in a conversation with a friend. Meanwhile, Alice, who works at the library, reorganizes shelf n and moves k as well as some other books to n'. She updates the library catalog and also leaves a sticky note at n indicating the new location of the moved books. Finally, Bob continues his way to n, reads the note, proceeds to n', and takes out k. The synchronization protocol of leaving a note (the *link*) when books are moved ensures that Bob can find k.

The library patron corresponds to a thread searching for and performing an operation on the key k stored at some node n in the tree having links and the librarian corresponds to a thread performing a split operation involving nodes n and n'. As in our library analogy, the synchronization technique of creating a forward pointer (the link) when nodes are split can work for various data organizations within and between nodes (e.g., if the nodes are organized as a B-tree or hash table). Hence, it applies to vastly different concrete data structures.

Our goal is to verify the correctness of *template algorithms* once and for all so that their proofs can be reused across different data structure implementations. The template algorithms act on an abstract model of data structures as acyclic graphs in which each edge is associated with a set of possible keys, e.g., in a binary search tree, the right-pointing child edge of a node with key 15 is associated with all key values greater than 15. To verify that a data structure implements a template algorithm, one then needs only to define how the concrete memory representation of the data structure is mapped to the abstract graph and check that this mapping meets all assumptions made by the template algorithm about how its primitive operations manipulate this graph. Crucially, this second verification step involves only sequential reasoning.

This idea of *algorithmic proof modularity* follows the classical approach of refinement-based reasoning. The technical challenge is to reconcile this idea with orthogonal techniques for achieving proof modularity, in particular, that of reasoning locally about modifications to the data structure graph, as it is supported, e.g., in separation logic (SL) [Ishtiaq and O'Hearn, 2001, O'Hearn et al., 2001, Reynolds, 2002]. Only by combining these orthogonal techniques can we obtain simple proofs that are easy to mechanize. As compared to the state of the art in verification technology, the proof of a template algorithm depends on certain invariants about the paths that a search for a key k follows in the data structure graph. In contrast, with the standard heap abstractions used in SL (e.g., inductive predicates), it is hard to express these synchronization invariants independently of the invariants that capture how the data structure is represented in memory. That is why existing proofs such as the one of the B-link tree in da Rocha Pinto et al. [2011] intertwine the synchronization invariants and the memory invariants, making the proof complex and difficult to reuse on different structures.

1.2 CONCURRENT SEARCH STRUCTURE TEMPLATES

This monograph shows how to adapt and combine recent advances in compositional abstractions and separation logic in order to achieve the envisioned algorithmic proof modularity for concurrent search structures.

We refer to the act of a thread searching for, inserting, or deleting a key k in a search structure as an operation on k, and to k as the *operation key*. We call the set of all keys the *key space* (e.g., the set of all natural numbers), written $\mathbb{K}$. For simplicity, the presentation in this monograph treats search structures as containing only keys, but all our proofs can be easily extended to consider search structures that store key-value pairs.

Our proof methodology for search structures is based on the template algorithms for concurrent search structures by Shasha and Goodman [1988], who identified the invariants needed for decoupling reasoning about synchronization and memory representation for such data structures. We apply those templates here for single-copy structures (structures containing at most one copy of a key at any point in time, for example B-trees), and extend them significantly to cover multicopy algorithms (structures that may contain multiple copies of a single key, for example log-structured merge trees [O'Neil et al., 1996]).

The second ingredient is the concurrent separation logic Iris [Jung et al., 2016, 2018, 2015, Krebbers et al., 2018, 2017a,b]. We show how to capture the high-level idea of our proof methodology in terms of new Iris resource algebras, yielding a general methodology for modular verification of concurrent search structures.

This methodology independently verifies (1) that each template algorithm satisfies the (atomic) abstract specification of search structures assuming that node-level operations maintain certain data structure-agnostic invariants and (2) that the implementations of these operations for each concrete data structure maintains these invariants.

The correctness criterion we target throughout this monograph is that of *linearizability* [Herlihy and Wing, 1990]. A concurrent data structure is linearizable if for every concurrent execution history of its operations, each operation appears to take effect at a single atomic step between its invocation and return point. We note that we focus on establishing partial correctness, which means that we do not prove progress guarantees, such as that a thread executing a search structure operation will always terminate. We discuss other correctness criteria as well as future extensions to proving progress properties in Chapter 14.

Our new resource algebras, in combination with Iris' notion of *atomic triples* [da Rocha Pinto et al., 2014, Jung et al., 2020, 2015], allows us to establish linearizability without having to explicitly reason about execution histories. Moreover, it yields a local proof technique that eliminates the need to reason explicitly about the global abstract state of the data structure. The latter crucially relies on the recently proposed *flow framework* [Krishna et al., 2018, 2020c], the final ingredient of our methodology.

The flow framework, as explained in Chapter 7, provides an SL-based abstraction mechanism that allows one to reason about global inductive invariants of general graphs in a local manner. Using this framework, we can do SL-style reasoning about the correctness of a concurrent search structure template while abstracting from the specific low-level heap representation of the underlying data structure.

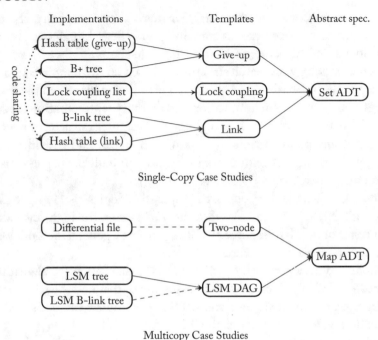

Figure 1.1: The decomposition of our proofs as a result of our template-based methodology. The dashed lines represent ongoing work.

1.3   CASE STUDIES

We demonstrate our methodology by mechanizing the correctness proofs of template algorithms for single-copy (Chapter 8) and multicopy (Chapter 9) search structures that abstract common real-world data structure implementations.

The single-copy templates we verify are based on the link, give-up, and lock-coupling technique of synchronization (Figure 1.1, top). For these, we derive concrete verified implementations based on B-trees, hash tables, and sorted linked lists, resulting in five different data structure implementations.

In the second half of the monograph, we apply the same approach to concurrent multicopy search structures. In such structures, inserts, deletes, and updates of a key k simply prepend a new instance of k to the structure with an appropriate value. For example, if key k has an associated value 5 and then this is changed to 17, an algorithm on a single-copy structure would find the node containing k and change the value from 5 to 17. On a multicopy structure, a new pair $(k, 17)$ would be prepended to the structure. A subsequent search on k would return the value associated with the most recent pair prepended to the structure, 17 in this example.

The pragmatic advantage of multicopy structures is that all modifications update the root node of the structure, so they can be processed very fast. The pragmatic disadvantage when compared to single-copy structures is that searches can take longer and the data structure may grow larger.

We verify a multicopy structure template called the LSM DAG template that generalizes the log-structured merge (LSM) tree to arbitrary DAG-like structures (Chapter 12). We derive an LSM tree implementation from the LSM DAG template (similar to what is seen in practice). The template also permits implementations that utilize DAG structures to organize the data, such as a data structure that combines the LSM tree and the B-link tree. In addition to the LSM DAG template, we also verify a two-node multicopy template that is motivated by the differential file data structure [Severance and Lohman, 1976].

A major advantage of our approach is that we can perform *sequential* reasoning when we verify that an implementation is a valid template instantiation. We therefore perform the template proofs in Iris/Coq and verify the implementations using the automated deductive verification tool GRASShopper [Piskac et al., 2013, 2014]. The automation provided by GRASShopper enables us to bring the proofs of highly complicated implementations such as B-link trees within reach.

Our proofs include a mechanization in both GRASShopper and Iris/Coq of the meta-theory of the flow framework. The verification efforts in each of the two systems are hence fully self-contained. The template proofs done in Iris are parametric with respect to any possible correct implementation of the node-level operations. The specifications assumed in Iris match those proved in GRASShopper. However, we note that there is no formal connection between the proofs done in the two systems. If one desires end-to-end certified implementations, one can perform both template and implementation proofs in Iris/Coq (albeit with substantial additional effort). Performing the proofs completely in GRASShopper or a similar SMT-based verification tool would require additional tooling effort to support reasoning about Iris-style resource algebras.

The proofs we obtain are more modular, reusable, and simpler than existing proofs of such concurrent data structures. Our experience is that adapting our technique to a new template algorithm and instantiating a template to a new data structure takes only a few hours of proof effort.

1.4 SUMMARY AND OUTLINE

This monograph describes a template-based methodology for verifying the safety of concurrent search structures. The methodology enables proofs to be compositional in terms of program structure and state, and to exploit thread and algorithmic modularity.

- Chapter 2 introduces some basic notation as well as the programming language we use throughout this monograph.

- All our proofs are performed in separation logic, in particular the higher-order concurrent separation logic Iris [Jung et al., 2018, 2015], which we introduce in Chapters 3 and 4.

- In Chapter 5, we define a novel resource algebra that allows us to use *ghost state* to keep track of *keysets*. The keyset of a node is the set of keys that are allowed in that node. *Keysets* let us reason locally about how changes to the contents of a node affect the global contents of the data structure.

- Chapter 6 presents our abstract graph model of search structures based on the edgeset framework. In particular, we define the keyset as an inductive quantity of the abstract search structure graph. This lets us separate the correctness of a search structure algorithm from the concrete memory representation of the data structure.

- Because the keyset is an inductive quantity of a graph, we use the flow framework (presented in Chapter 7) to reason locally about how changes to the structure of the data structure affect the keyset of a node.

- We demonstrate our proof technique on single-copy search structures in Chapter 8. We mechanically prove several complex real-world single-copy data structures such as the B-link tree and various hash structures. These case studies have appeared in preliminary form in a conference paper [Krishna et al., 2020b].

- We turn our attention to multicopy search structures in Chapter 9 and present an extension of the edgeset framework for such data structures in Chapter 10.

- The non-local linearization points of multicopy structure operations introduce additional proof complexity. In Chapter 11, we present a reusable proof to address this complexity.

- Chapter 12 presents a multicopy structure template algorithm that generalizes the LSM tree to structures based on arbitrary directed acyclic graphs.

- In Chapter 13, we discuss the mechanization of our proofs in Iris/Coq and GRASShopper. This effort also includes a mechanization of the meta-theory of the flow framework [Krishna et al., 2020c] in both tools, as well as a construction of a flow-interface resource algebra for proofs in Iris. These components are available online[1] and can be reused to support verification efforts on communication network graphs and memory management.

- In Chapter 14 we survey related work, discuss avenues for future work (such as liveness and lock-freedom), and conclude.

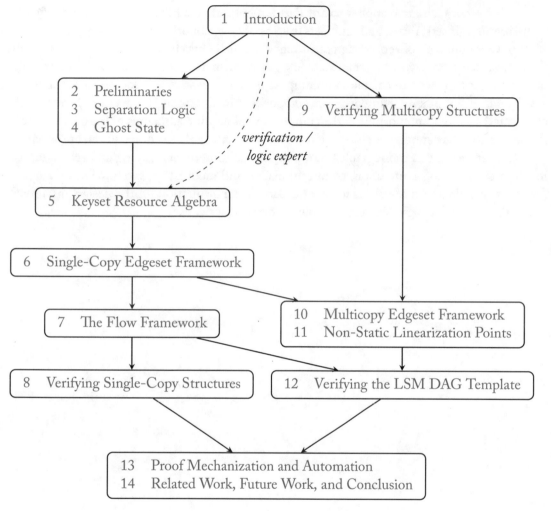

Figure 1.2: The chapter dependency graph.

Figure 1.2 shows the chapter dependency graph. The monograph is self-contained and aims to be also accessible to readers who have no prior background in formal verification. However, we note that a tutorial on the specific verification tools that we use for the proof mechanization, such as the Coq proof assistant, is outside the scope of this monograph. We provide references to further introductory materials and tutorials in §14.1 as well as in the accompanying online repository. Readers who are already familiar with Iris or a related concurrent separation logic, may want to skip the introductory chapters and continue directly with Chapter 5.

[1]https://github.com/nyu-acsys/template-proofs/tree/css_book

Our proof technique applies to any data structure that is indexed by keys, including implementations of sets, maps, and multisets (but, as of now, not other structures, e.g., queues and stacks). Our approach of separating concurrency templates and heap implementations requires the data structure to have an abstract state (e.g., as mathematical set or map) with a certain algebraic structure: we need to be able to decompose the abstract state into local abstract states that are *disjoint* in some sense. Moreover, the composition of abstract states needs to be associative, commutative, and homomorphic to the composition of disjoint subgraphs of the data structure.[2] All search structure implementations that we know of satisfy these compositional properties.

In summary, this monograph follows in the tradition of simplifying and scaling up verification efforts using abstraction, compositionality, and modularity. We hope our framework for verifying concurrent search structures can serve as an inspiration for the design and proof of many algorithms for high-performance multi-threaded systems.

---

[2]For instance, consider a binary search tree representing a mathematical map where each tree node stores a single key/value pair. If one arbitrarily splits the tree's graph into disjoint subgraphs, then these subgraphs represent disjoint mathematical maps whose union yields the map represented by the original composed heap graph.

CHAPTER 2

Preliminaries

This chapter provides technical background for some concepts used in this monograph, including basic mathematical notation, the programming language we use, and the concept of template algorithms using a single-node search structure template and implementation as an example.

2.1 BASICS AND NOTATION

We begin with some basic definitions and notation.

- The term $(b \; ? \; t_1 : t_2)$ denotes t_1 if condition b holds and t_2 otherwise.

- We write $f : A \to B$ for a total function from A to B, and $f : A \rightharpoonup B$ for a partial function from A to B.

- For a partial function f, we write $f(x) = \bot$ if f is undefined at x.

- We use the lambda notation $(\lambda x. \; E)$ to denote a function that maps x to the expression E (typically containing x).

- If f is a (partial) function from A to B, we write $f[x \rightarrowtail y]$ to denote the function from $A \cup \{x\}$ defined by $f[x \rightarrowtail y](z) := (z = x \; ? \; y : f(z))$.

- We use $\{x_1 \rightarrowtail y_1, \ldots, x_n \rightarrowtail y_n\}$ for pairwise different x_i to denote the function $\epsilon[x_1 \rightarrowtail y_1] \cdots [x_n \rightarrowtail y_n]$, where ϵ is the function on an empty domain.

- Given functions $f_1 : A_1 \to B$ and $f_2 : A_2 \to B$, we write $f_1 \uplus f_2$ for the function $f : A_1 \uplus A_2 \to B$ that maps $x \in A_1$ to $f_1(x)$ and $x \in A_2$ to $f_2(x)$ (if A_1 and A_2 are not disjoint sets, $f_1 \uplus f_2$ is undefined).

- We also write $\lambda_0 := (\lambda m. \; 0)$ for the identically zero function and $\lambda_{id} := (\lambda m. \; m)$ for the identity function.

- For functions f_1, f_2, we write $f_2 \circ f_1$ to denote function composition, i.e., $(f_2 \circ f_1)(x) = f_2(f_1(x))$, and use superscript notation f^p to denote the function composition of f with itself p times.

- For multisets, we use the standard set notation if it is clear from the context. We also write $\{x_1 \rightarrowtail i_1, \ldots, x_n \rightarrowtail i_n\}$ for the multiset containing i_1 occurrences of x_1, i_2 occurrences of x_2, etc. For a multiset S, we write $S(x)$ to denote the number of occurrences of x in S.

- We write _ for an anonymous variable, usually to bind a variable that is never used.

We now turn to introducing some basic algebraic concepts that will be used in Chapters 4 and 7.

Definition 2.1   A *partial monoid* is a set M, along with a partial binary operation $+: M \times M \rightharpoonup M$, and a special zero element $0 \in M$, such that (1) $+$ is associative, i.e., $(m_1 + m_2) + m_3 = m_1 + (m_2 + m_3)$; and (2) 0 is an identity element, i.e., $m + 0 = 0 + m = m$. Here, equality means that either both sides are defined and equal, or both sides are undefined.

Partial monoids are the basis of ghost state, an important reasoning technique which will be introduced in Chapter 4. An example of a partial monoid is the set $\mathcal{P}(S)$ of all subsets of an arbitrary set S, together with disjoint union $\uplus$ (where $S_1 \uplus S_2$ is undefined if they are not disjoint) and the empty set $\emptyset$. We usually identify a partial monoid $(M, +, 0)$ with its support set M.

If $+$ is a total function, then we call M a monoid. For example, the set of natural numbers $\mathbb{N}$ together with addition $+$ and zero form a monoid. Let $m_1, m_2, m_3 \in M$ be arbitrary elements of the (partial) monoid in the following. Here are some terminology and notation associated with (partial) monoids:

- We call a (partial) monoid M *commutative* if $+$ is commutative, i.e., $m_1 + m_2 = m_2 + m_1$. $\mathbb{N}$ and $\mathbb{Z}$ are commutative monoids, while 2×2 matrices of natural numbers with matrix multiplication and the identity matrix form a non-commutative monoid.

- Similarly, a commutative (partial) monoid M is *cancellative* if $+$ is cancellative, i.e., if $m_1 + m_2 = m_1 + m_3$ is defined, then $m_2 = m_3$. For example, $\mathbb{N}$ is cancellative monoid while the monoid formed by sets of natural numbers under set union is not (as $\{1\} \cup \{1, 2\} = \{1\} \cup \{2\}$).

- We say M is *positive* if $m_1 + m_2 = 0$ implies that $m_1 = m_2 = 0$. As you might expect, $\mathbb{N}$ is a positive (partial) monoid, while $\mathbb{Z}$ is not positive.

- For a positive (partial) monoid M, we can define a partial order $\leq$ on its elements as $m_1 \leq m_2$ if and only if $\exists m_3. \ m_1 + m_3 = m_2$. Positivity also implies that every $m \in M$ satisfies $0 \leq m$. For $\mathbb{N}$, this order corresponds to the natural less-than-or-equal-to ordering on natural numbers.

We will see commutative cancellative monoids used in Chapter 7 in order to set up the flow framework for reasoning locally about global graph properties.

2.2   PROGRAMMING LANGUAGE

The programming language that we use in this monograph is an ML-like language with higher-order store, fork, and compare-and-set (**CAS**, also sometimes called compare-and-swap), whose grammar is given below:

$e \in$ Expr ::=	x	Variable
|	$()$	Unit constant
|	z	Integer constants
|	`true` | `false`	Boolean constants
|	$e_1 + e_2$ | $e_1 - e_2$ | $\ldots$	Arithmetic expressions
|	$e_1 == e_2$ | $e_1 <= e_2$ | $\ldots$	Boolean expressions
|	(e_1, e_2)	Pair (tuple) expressions
|	**if** e_1 **then** e_2 **else** e_3	Conditional
|	$e_1\ e_2$	Function application
|	$(\mu\ f\ x.\ e)$	(Recursive) function
|	**let** $x = e_1$ **in** e_2	Let binding
|	$\text{Inj}_1\ e$ | $\text{Inj}_2\ e$	Constructor expressions
|	**match** e **with** $\text{Inj}_1\ x \rightarrow e_1$ | $\text{Inj}_2\ x \rightarrow e_2$	Pattern matching
|	**ref**(e)	Reference creation
|	$!e$	Dereference
|	$e_1 \leftarrow e_2$	Reference assignment
|	**CAS**(e, e_1, e_2)	Compare-and-set
|	**fork** $\{e\}$	Fork

As is standard in languages based on the λ-calculus, all programs are expressions, the simplest kind of which is just a variable or constant value. The constant values in our language include the unit value (the return value of expressions that are evaluated only for their side-effects such as assignments), integers, and Booleans. Arithmetic and Boolean expressions use infix notation and conditional expressions are as expected.

Function application is written in standard functional programming style as `foo arg` instead of `foo(arg)`. It is also left-associative, i.e., a nested function application `foo arg1 arg2` should be read as `(foo arg1) arg2`. A function expression $(\mu\ f\ x.\ e)$ evaluates to the function that satisfies the (possibly recursive) equation $f(x) = e$. For example, the following expression evaluates to the factorial function

$$(\mu\ \mathit{fac}\ x.\ \textbf{if}\ x == 0\ \textbf{then}\ 1\ \textbf{else}\ x * \mathit{fac}\ (x - 1))$$

A lambda abstraction expression $(\lambda x.\ e)$, which describes a function that maps argument x to the expression e, can be defined as a syntactic shorthand $(\lambda x.\ e) := (\mu\ \_\ x.\ e)$.

A let binding expression **let** $x = e_1$ **in** e_2 binds the variable x to the result value of the expression e_1, and evaluates e_2 with this new binding. That is, e_2 in general would depend on x, so is evaluated assuming x is equal to e_1. The value obtained from e_2 is then also the result value of the whole let binding expression. In the rest of the monograph, we use standard syntactic shorthands, such as:

$$e_1; e_2 \quad := \quad \textbf{let}\ \_ = e_1\ \textbf{in}\ e_2$$
$$\textbf{let}\ f\ x = e_1\ \textbf{in}\ e_2 \quad := \quad \textbf{let}\ f = (\lambda x.\ e_1)\ \textbf{in}\ e_2$$
$$\textbf{let rec}\ f\ x = e_1\ \textbf{in}\ e_2 \quad := \quad \textbf{let}\ f = (\mu\ f\ x.\ e_1)\ \textbf{in}\ e_2$$

We also omit the **in** keyword when defining top-level functions.

The language provides constructors Inj_1 and Inj_2 to construct values of a generic disjoint sum type. If e evaluates to value v, then $\text{Inj}_i\ e$ evaluates to $\text{Inj}_i\ v$ which can be thought of as a pair consisting of v and a tag bit that encodes i. Such tagged values can be decomposed using pattern matching expressions such as:

$$\textbf{match}\ e_0\ \textbf{with}\ \text{Inj}_1\ x \mathrel{->} e_1 \mid \text{Inj}_2\ x \mathrel{->} e_2$$

Here, if e_0 evaluates to $\text{Inj}_i\ v$, then the match expression binds x to v and continues evaluating e_i under this binding. The result value of the match expression is that of the chosen e_i. Algebraic data types, which are commonly supported in functional languages, can be encoded using disjoint sums. For instance, consider the ML type α option where α is a type parameter. This type has two constructors: None and Some e where e must evaluate to a value of type α. The type can be used, e.g., to turn a partially defined function returning values of type α into a total function returning values of type α option by using None to indicate the absence of a return value and Some x to indicate that the return value is x. A caller of the function can then distinguish these two cases by pattern matching on the constructor of the return value. We can encode this type by letting None $:= \text{Inj}_1\ ()$ and Some $x := \text{Inj}_2\ x$.

The reference creation expression $\textbf{ref}(e)$ evaluates to the address of a newly allocated heap location, whose value is set to the result of evaluating the expression e. Heap locations, or references, can be read (or *dereferenced*) by using the $!e$ command, where e evaluates to a heap location. A heap location obtained by evaluating an expression e_1 can be updated to the value of e_2 using an assignment $e_1 \leftarrow e_2$, which returns the unit value $()$. The compare-and-set expression $\textbf{CAS}(e, e_1, e_2)$ is an instruction that checks if the value at heap location ℓ obtained by evaluating e is equal to the value of e_1; if so, it sets ℓ's value to the value of e_2 and returns true, otherwise it does nothing and returns false. The compare-and-set is *atomic* once the argument expressions are evaluated, which means that during the execution of the comparison and update, no other thread can read from or write to ℓ (in other words, they appear to take place instantaneously). The fork command $\textbf{fork}\ \{e\}$ creates a new thread that evaluates expression e in parallel with the forking thread. The expression returns the unit value $()$ without waiting for the new thread to complete its execution.

```
 1 let create () =                            1 let allocRoot _ =
 2   let r = allocRoot () in                   2   ref(⊥)
 3   r                                         3
 4                                             4 let decisiveOp ω r k =
 5 let cssOp ω r k =                           5   match ω with
 6   lockNode r;                               6   | search -> search r k
 7   let res = decisiveOp ω r k in             7   | insert -> insert r k
 8   unlockNode r;                             8   | delete -> delete r k
 9   res                                       9
10                                            10 let rec insert r k =
11 let rec lockNode x =                       11   let u = !r in
12   if CAS(lk(x), false, true) then          12   if u == ⊥ then
13     ()                                      13     r ← k;
14   else                                      14     true
15     lockNode x                              15   else
16                                             16     insert r k
17 let unlockNode x =
18   lk(x) ← false
```

Figure 2.1: A template algorithm for a single-node search structure (left), and a simplistic example of an implementation for the single-node template (right). Template algorithms use helper functions (in purple, e.g., decisiveOp) to be implementation-agnostic; their definitions are provided by each implementation.

2.3 TEMPLATE ALGORITHMS

In this section we present an example of a template algorithm for the simplest possible search structure, a single-node structure (Figure 2.1, left). The create function creates a new search structure, by calling a allocRoot helper function, and returns the address of the root of the newly created structure.

The cssOp function (for concurrent search structure operation) stands for any one of the three core search structure operations, by means of the parameter ω. Recall that search structures are data structures implementing a mathematical set data type, so ω is either search, insert, or delete. cssOp first locks the (only) node r, then calls a function decisiveOp on the locked node, before unlocking the node and returning the result. The Boolean value returned indicates if the operation modified the search structure. For example, an insertion returns true if the given key was not already present in the structure.

For simplicity of exposition, we present a spinlock implementation of lockNode and unlockNode. We also assume that there exists a function lk(x) that maps each node x to the heap address that it uses as a lock flag (true when locked, false otherwise). x is locked by repeatedly trying to **CAS** lk(x) to true, which will succeed only if the node is unlocked. Unlocking is simpler, and the thread that has locked a node sets lk(x) to false.

We call this algorithm a template algorithm because it does not specify the implementation of its helper functions (in this case, the functions `allocRoot` and `decisiveOp`), denoted in purple throughout this monograph. The function `decisiveOp` is the *decisive operation* that performs the actual insertion, removal, or search on the node. Since the implementation of `decisiveOp` depends on the concrete implementation of the data structure (how nodes are laid out in memory, or how keys are stored within nodes), we abstract from it by making it a helper function.

Figure 2.1 (right) shows an implementation of the single-node template. The simplest implementation is for the node r to consist of a single heap location (also, for convenience, at address r). The implementation of `allocRoot` creates this heap cell and returns its address. The implementation of insert reads the address r into variable u and if u is empty (denoted by some default initial value $\bot$), writes the operation key k to location r. However, if u is not empty, then this simplistic implementation loops infinitely. While not necessary for partial correctness, one could modify the implementation to return false if u were equal to k and loop otherwise. A more realistic implementation might implement the single node as an array of keys, and dynamically resize it when full.

We will show that the correctness of template algorithms does not depend on the exact implementations of helper functions such as `decisiveOp`. Instead, we specify the expected behavior of helper functions in an implementation-agnostic way. For example, helper function `decisiveOp` is required to correctly perform its operation on a single node but our specification does not require the node's representation to be either the single-heap-cell implementation or the array implementation described above. We will see in §3.4 how to specify `decisiveOp` for the single-node template, and in Chapter 8 how to specify it for search structures consisting of multiple nodes.

CHAPTER 3

Separation Logic

In this monograph, we use *separation logic* to specify and verify concurrent data structures. Separation logic (SL) is an extension of Hoare logic [Hoare, 1969] that is tailored to perform modular reasoning about programs that manipulate mutable resources. In other words, SL is a language that allows one to succinctly and modularly describe states of a program. Each sentence in this language is called a *proposition* (or *formula*, *assertion*), and describes a set of states. We say a state *satisfies* a proposition when the state is described by the proposition. Separation logic also gives us a set of proof rules that can be used to prove that states of interest (such as the set of resulting states after a program executes) satisfy a particular proposition.

There are many incarnations of SL, each tailored to reasoning about a particular class of programs. In this monograph, we use *Iris*, a mechanized higher-order concurrent separation logic framework. The distinguishing feature of Iris is its generality: it is designed as a small set of core primitives and proof rules that can be used to encode a large variety of common constructs and techniques for reasoning about concurrent programs. In particular, Iris is easily extendable with user-defined resources via its ghost state mechanism, which we will describe in Chapter 4.

The price paid for this generality is that the core Iris logic is very abstract. To make this monograph accessible to a wider audience, we present many of the derived features as though they are primitive Iris features, and avoid talking about the formal semantic model altogether. We refer the interested reader to a paper by Jung et al. [2018] for a more detailed introduction to Iris, the Iris tutorial at POPL 2021 [Chajed et al., 2021] for a gentle introduction, and to the documentation [Iris Team, 2020] for the full details. We discuss other SLs, as well as alternate approaches to verifying data structures, in Chapter 14.

Note that Iris is a garbage-collected or *intuitioniztic* separation logic, hence all programs in this monograph assume a garbage-collected setting. We can extend the techniques presented in this monograph to also prove absence of memory leaks by using an extension of Iris such as Iron [Bizjak et al., 2019].

3.1 SEPARATION LOGIC BY EXAMPLE

Consider the following program expression that reads the value stored at heap location x into a variable ($v = !x$) and then writes $v + 1$ back into the location:

$$e_{\mathsf{inc}} := \mathbf{let}\ v = !x\ \mathbf{in}\ x \leftarrow (v + 1)$$

Informally, e_{inc} is a program that increments the value stored at the heap location x.

We can formalize this specification in SL by using a *Hoare triple* [Hoare, 1969], which is an assertion of the form $\{P\}\ e\ \{v.\ Q\}$, where P and Q are propositions. $\{P\}\ e\ \{v.\ Q\}$ is true if for every state σ that satisfies P we have that (1) the program e does not reach an error state when run from σ (for example, by trying to read unallocated memory), and (2) that if e terminates then it returns some value v and the new state is some σ' that satisfies Q. We call P the *precondition* and Q the *postcondition* of e, and we write $\{P\}\ e\ \{Q\}$ in the case where Q does not mention the return value v.

Here is the desired specification of our example program e_{inc}:[1]

$$\{x \mapsto n\}\ e_{\mathsf{inc}}\ \{x \mapsto n + 1\}$$

The precondition here uses a *points-to* predicate $x \mapsto n$, a primitive proposition asserting that the program state contains a heap cell at address x containing value n. The postcondition uses a similar points-to predicate, and, in simple words, the triple says: if the program e_{inc} is run on a state that contains a heap cell at address x with value n, then it results in a state where the cell x contains $n + 1$. It also implicitly asserts that e_{inc} does not crash, by, e.g., attempting to dereference memory that is not allocated or divide by zero.

Before we can prove that e_{inc} meets its specification, we need to introduce some proof rules for Hoare triples (Figure 3.1). A proof rule, or inference rule, consists of two parts separated by a horizontal line: the part above the line contains one or more premises, and the part below the line contains the conclusion. For example, the HOARE-DISJ rules tells us that if we can prove the two triples $\{P\}\ e\ \{v.\ R\}$ and $\{Q\}\ e\ \{v.\ R\}$, then we can apply the rule and infer that e also satisfies the triple $\{P \vee Q\}\ e\ \{v.\ R\}$. A rule with no premises is called an *axiom*, and in this case we omit the horizontal line (e.g., HOARE-LOAD). Some rules are bi-directional: the premise implies the conclusion, but the conclusion also implies the premise. These rules are denoted with a double horizontal line (e.g., HOARE-EXIST). We will explain the proof rules in Figure 3.1 as and when we use them to prove our example program e_{inc}, and explain the remaining rules at the end of this section.

Many classical proof systems, including the Coq proof assistant, usually perform what is known as *backward reasoning*. In this style of reasoning, we start with a specification that we want to prove (the *goal*), and then find an inference rule that matches the structure of the goal and apply it to replace the goal with the premises of the rule.

Here is how we can prove this specification using the proof rules in Figure 3.1:

$$\frac{\dfrac{\dfrac{\rule{3cm}{0.4pt}}{\{x \mapsto n\}\ x \leftarrow (v + 1)\ \{x \mapsto v + 1\}}\ \text{HOARE-STORE}}{\dfrac{\{x \mapsto n\ *\ v = n\}\ x \leftarrow (v + 1)\ \{x \mapsto v + 1\ *\ v = n\}}{\forall v.\ \{x \mapsto n\ *\ v = n\}\ x \leftarrow (v + 1)\ \{x \mapsto n + 1\}}\ \text{HOARE-CSQ}}\ \text{HOARE-FRAME}\qquad \dfrac{}{\{x \mapsto n\}\ !x\ \{v.\ x \mapsto n\ *\ v = n\}}\ \text{HOARE-LOAD}}{\{x \mapsto n\}\ \mathbf{let}\ v\ =\ !x\ \mathbf{in}\ x \leftarrow (v + 1)\ \{x \mapsto n + 1\}}\ \text{HOARE-LET}$$

[1]In all such triples that we use as specifications in this monograph, free variables such as n are implicitly universally quantified.

HOARE-RET
$$\{\mathsf{True}\}\; w\; \{v.\, v = w\}$$

HOARE-FALSE
$$\{\mathsf{False}\}\; e\; \{v.\, P\}$$

HOARE-ALLOC
$$\{\mathsf{True}\}\; \mathbf{ref}(v)\; \{\ell.\, \ell \mapsto v\}$$

HOARE-LOAD
$$\{\ell \mapsto v\}\; {!}v\; \{w.\, \ell \mapsto v \;*\; w = v\}$$

HOARE-CAS-SUC
$$\{\ell \mapsto v\}\; \mathbf{CAS}(\ell, v, w)\; \{b.\, b = \mathit{true} \;*\; \ell \mapsto w\}$$

HOARE-STORE
$$\{\ell \mapsto v\}\; \ell \leftarrow w\; \{\ell \mapsto w\}$$

HOARE-CAS-FAIL
$$\frac{v \neq v'}{\{\ell \mapsto v\}\; \mathbf{CAS}(\ell, v', w)\; \{b.\, b = \mathit{false} \;*\; \ell \mapsto v\}}$$

HOARE-LAM
$$\frac{\{P\}\; e[x \rightarrowtail v]\; \{w.\, Q\}}{\{P\}\; (\lambda x.\, e)\, v\; \{w.\, Q\}}$$

HOARE-FORK
$$\frac{\{P\}\; e\; \{\mathsf{True}\}}{\{P\}\; \mathbf{fork}\,\{e\}\; \{\mathsf{True}\}}$$

HOARE-LET
$$\frac{\{P\}\; e_1\; \{w.\, R\} \qquad \forall w.\, \{R\}\; e_2[x \rightarrowtail w]\; \{v.\, Q\}}{\{P\}\; \mathbf{let}\; x = e_1 \;\mathbf{in}\; e_2\; \{v.\, Q\}}$$

HOARE-CSQ
$$\frac{P \Rightarrow P' \qquad \{P'\}\; e\; \{v.\, Q'\} \qquad \forall v.\, Q' \Rightarrow Q}{\{P\}\; e\; \{v.\, Q\}}$$

HOARE-FRAME
$$\frac{\{P\}\; e\; \{v.\, Q\}}{\{P * R\}\; e\; \{v.\, Q * R\}}$$

HOARE-DISJ
$$\frac{\{P\}\; e\; \{v.\, R\} \qquad \{Q\}\; e\; \{v.\, R\}}{\{P \vee Q\}\; e\; \{v.\, R\}}$$

HOARE-EXIST
$$\frac{\forall x.\, \{P\}\; e\; \{v.\, Q\}}{\{\exists x.\, P\}\; e\; \{v.\, Q\}}$$

Figure 3.1: Proof rules for establishing Hoare triples. We write $e[x \rightarrowtail v]$ to denote the expression e after substituting all occurrences of the variable x with the term v.

When explaining a proof using backward reasoning, we read these *proof trees* starting from the goal at the bottom (the root of the tree) and work our way up. Each horizontal line represents the use of an inference rule to transform the current goal into one or more (hopefully simpler) goals.

For example, the first rule we use in this proof is HOARE-LET, which works on let-bindings and breaks up the proof into, in our case, the proof of the bound expression $!x$ (on the left), and the proof of the let-body $x \leftarrow (v + 1)$ (on the right). Note that HOARE-LET requires the user to come up with an intermediate proposition R that describes the states of the program after evaluating the bound expression e_1 but before evaluating the let-body e_2. (R also usually specifies the value that the bound expression evaluated to, in our case it will say that v is the value stored at x.) The most common way to construct R is to defer the choice until we finish the proof of the left-hand side branch, as this will give us a clue as to what R should be. In this case, since e_1 is a dereference, our only option is to use the rule HOARE-LOAD, but note that this tells us exactly what R should be.

The postcondition of $!x$ is the proposition $x \mapsto n * v = n$, which uses the *separating conjunction* connective $*$. We will explain $*$ in detail below (§3.2), but for now it is enough to think of it as standard logical conjunction $\wedge$ and read $x \mapsto n * v = n$ as describing a state both containing the heap cell $x \mapsto n$ and satisfying $v = n$.

The proof tree that we have constructed so far looks like:

$$\frac{\dfrac{}{\{x \mapsto n\}\, !x\, \{v.\, x \mapsto n * v = n\}} \text{ HOARE-LOAD} \qquad \forall v.\, \{x \mapsto n * v = n\}\, x \leftarrow (v+1)\, \{x \mapsto n + 1\}}{\{x \mapsto n\}\, \mathbf{let}\ v\ =\ !x\ \mathbf{in}\ x \leftarrow (v+1)\, \{x \mapsto n + 1\}} \text{ HOARE-LET}$$

Our goal is now to complete the right branch, where we have to prove

$$\forall v.\, \{x \mapsto n * v = n\}\, x \leftarrow (v+1)\, \{x \mapsto n + 1\}.$$

Again, looking at the program expression, our only option is to use HOARE-STORE, but the current proof goal does not quite fit. In particular, HOARE-STORE will give us the postcondition $x \mapsto v + 1$ while we need the postcondition $x \mapsto n + 1$. To infer this, we need the fact that $v = n$, which is something that we know in the precondition of the current goal. If we can somehow transfer the fact $v = n$ from the precondition to the postcondition of the program fragment $x \leftarrow (v+1)$, then we can use HOARE-CSQ to complete the proof tree:[2]

$$\frac{\dfrac{}{\{x \mapsto n\}\, !x\, \{v.\, x \mapsto n * v = n\}} \text{ HOARE-LOAD} \qquad \dfrac{\dfrac{\dfrac{\dfrac{}{\{x \mapsto n\}\, x \leftarrow (v+1)\, \{x \mapsto v + 1\}} \text{ HOARE-STORE}}{\{x \mapsto n * v = n\}\, x \leftarrow (v+1)\, \{x \mapsto v + 1 * v = n\}} \text{ HOARE-FRAME}}{\forall v.\, \{x \mapsto n * v = n\}\, x \leftarrow (v+1)\, \{x \mapsto n + 1\}} \text{ HOARE-CSQ}}{\{x \mapsto n\}\, \mathbf{let}\ v\ =\ !x\ \mathbf{in}\ x \leftarrow (v+1)\, \{x \mapsto n + 1\}}} \text{ HOARE-LET}$$

As we will explain in more detail later, every proposition in SL can be thought of as a *resource*. It may be natural to think of the predicate $x \mapsto n$ as a resource, as it describes a heap cell or a part of the program state, but while performing SL proofs we also think of facts or bits of knowledge, such as $v = n$, as resources. The *frame rule* HOARE-FRAME lets us *frame* resources around a program fragment. This means we can carry resources that are not needed by the program fragment from its precondition to its postcondition. In our case, the program fragment is $x \leftarrow (v+1)$, and as this does not modify v or n, we frame the resource $v = n$, allowing us to use HOARE-STORE to finish the proof.

Despite constructing this proof tree using backward reasoning, at the point when we had two goals to prove, we found it more natural to complete the one that was earlier in program

---

[2]The $\Rrightarrow$ connective used in HOARE-CSQ is the *view shift* operator from Iris. For now, we can think of it as implication, but it also allows us to perform updates on ghost state (see Chapter 4).

order (the left subtree). In most cases, the structure of the current program expression determines the proof rule we must use, as well as any intermediate proposition that links the first goal to the next one. To reflect this, we rotate the proof tree by 90 degrees to align the proof order with program order:[3]

$$\{x \mapsto n\}$$

$$\text{\textbf{let }} v \ =$$
$$\{x \mapsto n\}$$

HOARE-LET · HOARE-LOAD

$$!x$$

$$\{v. \, x \mapsto n \, * \, v = n\}$$
$$\text{\textbf{in}}$$
$$\{x \mapsto n \, * \, v = n\}$$
$$\{x \mapsto n\}$$

HOARE-FRAME · HOARE-STORE

$$x \leftarrow (v \ + \ 1)$$

$$\{x \mapsto v + 1\}$$
$$\{x \mapsto v + 1 \, * \, v = n\}$$
$$\{x \mapsto n + 1\}$$

Since all the proofs in this monograph are constructed in this way, i.e., by selecting the proof rule associated with each program construct, we write the same proof tree in a more compact, inline style as follows:

```
1 {x ↦ n}
2 let v = !x in (* HOARE-LOAD *)
3 {x ↦ n * v = n}
4 x ← (v + 1) (* HOARE-FRAME and HOARE-STORE *)
5 {x ↦ v + 1 * v = n}
6 {x ↦ n + 1} (* HOARE-CSQ *)
```

The way to read such an inline-style proof is that the first and last line specify the Hoare triple for the entire program: $\{x \mapsto n\} \ e_{\mathsf{inc}} \ \{x \mapsto n + 1\}$. Every alternate line is an intermediate proposition describing the states of the program at that point. For example, the intermediate proposition $\{x \mapsto n \, * \, v = n\}$ is both the postcondition of the first line of code as well as the precondition of the second line of code. Note that the program expression in each line of code determines the proof rule that must be used to prove the triple for that line. In addition, we omit standard

---

[3]As you can imagine, upright proof trees quickly take up too much horizontal space when reasoning about realistic programs.

rules like HOARE-FRAME and HOARE-CSQ when it is clear how to apply them. In more complex scenarios, we use comments or multiple intermediate propositions to clarify how to get from one proof step to the next (for example, the last two lines of the proof above). We call the intermediate propositions in the inline-style proofs our "proof context" at that point, because this is the set of all resources that are available at that point in the program.

Other Hoare Proof Rules.   We have seen the proof rules HOARE-LET, HOARE-LOAD, HOARE-CSQ, HOARE-FRAME, and HOARE-STORE in action. The other rules in Figure 3.1 are also fairly straightforward, such as HOARE-RET for dealing with return values, HOARE-LAM for lambda expressions, and HOARE-ALLOC for allocating a new heap cell in memory. HOARE-FALSE allows one to prove anything if the precondition is invalid. HOARE-DISJ and HOARE-EXIST allow one to reason about Boolean connectives and quantifiers. We defer the discussion of HOARE-FORK, HOARE-CAS-SUC, and HOARE-CAS-FAIL to §3.2 where we discuss concurrent programs.

3.2   SEPARATING CONJUNCTION

Consider the following program that writes to two heap cells in two parallel threads:

$$e_{\mathsf{par}} := (x \leftarrow 1 \parallel y \leftarrow 2)$$

In the above program we use the commonly used parallel composition syntax $(e_1 \parallel e_2)$ which denotes forking two threads that run e_1 and e_2, respectively, and waiting until they complete. Parallel composition can be defined in terms of **fork** and **CAS**. Note that this program is safe only if x and y are *distinct* heap locations; if they are equal, then the program has undefined behavior as the two threads would race to write to the same heap cell.

To specify this program, we use the separating conjunction $*$. Unlike standard conjunction $\wedge$, separating conjunction conjoins two propositions that describe *disjoint* portions of program state. For instance, $x \mapsto \_ \wedge y \mapsto \_$ asserts that x is a heap cell and y is a heap cell (but they could be the same heap cell), while $x \mapsto \_ * y \mapsto \_$ asserts that x is a heap cell *and separately* y is a heap cell.

Formally, a state σ satisfies $P * Q$ if it can be broken up into two disjoint states $\sigma = \sigma_1 \odot \sigma_2$ such that σ_1 satisfies P and σ_2 satisfies Q. When P and Q are propositions denoting heaps, this means that they talk about disjoint regions of the heap, i.e., that they do not have any heap addresses in common. In particular, this means that $x \mapsto \_ * y \mapsto \_$ implies that $x \neq y$, while the proposition $x \mapsto y * x \mapsto z$ is unsatisfiable as it is not possible to split a heap into two disjoint portions both of which contain the same address x.

We can thus specify e_{par} as follows:

$$\{x \mapsto \_ * y \mapsto \_\} \, (x \leftarrow 1 \parallel y \leftarrow 2) \, \{x \mapsto 1 * y \mapsto 2\}$$

To prove this specification, we use the following parallel composition rule (which can be derived from HOARE-FORK):

HOARE-PAR

$$\frac{\{P_1\}\, e_1\, \{Q_1\} \qquad \{P_2\}\, e_2\, \{Q_2\}}{\{P_1 * P_2\}\, (e_1 \parallel e_2)\, \{Q_1 * Q_2\}}$$

HOARE-PAR tells us that executing two threads in parallel is safe if both threads operate on disjoint portions of the state. We can use HOARE-FORK to verify e_{par} as follows:

$$\{x \mapsto \_ \ast y \mapsto \_\}$$
$$\{x \mapsto \_\} \quad \Big\| \quad \{y \mapsto \_\}$$
$$x \leftarrow 1 \quad \Big\| \quad y \leftarrow 2$$
$$\{x \mapsto 1\} \quad \Big\| \quad \{y \mapsto 2\}$$
$$\{x \mapsto 1 \ast y \mapsto 2\}$$

Returning to our analogy of thinking of SL propositions as resources, in the concurrent setting we can think of each thread *owning* the resources in its precondition. HOARE-PAR then tells us that a thread can subdivide its resources and distribute them among the subthreads that it spawns. Primitive resources such as points-to predicates $x \mapsto \_$ are not subdividable, so there is no way to use HOARE-PAR alone to prove that the program $(x \leftarrow 1 \parallel x \leftarrow 2)$ is safe. In such situations, we think of resources such as $x \mapsto \_$ as being exclusively owned by a single thread.

Given this, it may seem counter-intuitive that the rules HOARE-CAS-SUC and HOARE-CAS-FAIL require exclusive ownership of the location ℓ despite the fact that the main application of CAS is fine-grained concurrent programs where multiple threads access the same heap cell. However, we will see in §3.7 proof rules for atomic commands such as CAS (compare-and-set) that allow one to take exclusive ownership of shared resources for the duration of an atomic command (since no other thread can interfere).

3.3 PROPOSITIONS

Iris propositions describe the *resources* owned by a thread. These resources can be part of a concrete program state, for example, a set of heap cells, which captures the situations in which a thread has exclusive ownership over these cells (because, say, it has locked them). To reason about fine-grained concurrency and more complex concurrency patterns, these resources can also capture shared ownership and partial knowledge of shared program state. In order to build intuition, we focus on the simple case where propositions describe concrete program states, in the form of subsets of the heap, and defer the discussion of advanced resources to Chapter 4.

For a formal definition of program states and the satisfaction relation in Iris, see the Iris documentation [Iris Team, 2020].

The grammar of the subset of Iris propositions that we use throughout the monograph is shown in Figure 3.2, and includes the following constructs:

$$P, Q, R ::= \text{True} \mid \text{False} \mid P \wedge Q \mid P \vee Q \mid P \Rightarrow Q$$
$$\mid \exists x.\ P \mid \forall x.\ P$$
$$\mid x \mapsto v \mid P * Q \mid P \mathbin{-\!\!*} Q \mid \textstyle\bigstar_{x \in X} P$$
$$\mid \boxed{P}^{\mathcal{N}} \mid \triangleright P \mid \lceil a \rceil^{\gamma} \mid P \Rrightarrow Q$$
$$\mid \{P\}\ e\ \{v.\ Q\} \mid \langle x.\ P \rangle\ e\ \langle v.\ Q \rangle \mid \text{AU}_{x.P,Q}(\Phi)$$

Figure 3.2: The grammar of Iris propositions used in this monograph.

- The first line consists of standard propositional constructs: the propositions True and False[4], then conjunction, disjunction, and implication.

- The second line introduces quantification. Note that since Iris is a *higher-order* logic, x can range over any type, including that of propositions and (higher-order) predicates.

- We have already seen the points-to proposition and separating conjunction on the third line. Note that Iris is an affine logic, which means $P * Q \vdash P$[5] holds for any propositions P and Q. In particular, this implies that if a state σ satisfies $x \mapsto v$, then σ can possibly contain more than just the heap cell at address x.

- We also have an iterated version of separating conjunction: $\bigstar_{x \in X} P$, where the bound variable x ranges over a finite set X. For example, $\bigstar_{x \in X} x \mapsto 13$ denotes states that contain at least the set of heap cells with addresses in X all of whom store the value 13.

- The *separating implication* connective $\mathbin{-\!\!*}$, also known as the *magic wand*, is defined as: a state σ satisfies $P \mathbin{-\!\!*} Q$ if for every state σ_1 disjoint from σ that satisfies P, the combined state $\sigma \odot \sigma_1$ satisfies Q. The best way to understand $\mathbin{-\!\!*}$ is to think of $*$ and $\mathbin{-\!\!*}$ as separation logic analogues of $\wedge$ and $\Rightarrow$ from first-order logic. For instance, $P * Q$ means you have both P *and* Q (and that they are disjoint, or more generally, composable). Similarly, $P \mathbin{-\!\!*} Q$ describes a state such that if you conjoin it with a state satisfying P, then you get a state satisfying Q. This property, $P * (P \mathbin{-\!\!*} Q) \vdash Q$, is the SL analog of *modus ponens* in first-order logic ($P \wedge (P \Rightarrow Q) \vdash Q$).

- The fourth line contains the invariant proposition $\boxed{P}^{\mathcal{N}}$ and later modality $\triangleright P$ (both explained in §3.5), as well as the ghost state proposition $\lceil a \rceil^{\gamma}$ and the view shift $P \Rrightarrow Q$ (both explained in Chapter 4).

- Finally, the last line contains Hoare and atomic triples (explained in §3.6).

---

[4]The propositions True and False are semantically distinct from the Boolean constants *true* and *false* which are values of our programming language. Intuitively, propositions can be thought of as describing sets of program states, where a program state maps program variables to values.

[5]$\vdash$ is the *provable entailment* relation; $P \vdash Q$ means if we can prove P is true, then we can prove that Q is true.

3.4 ABSTRACT PREDICATES

In this section, we introduce *abstract predicates*, a powerful tool for modular proofs in separation logics.

Recall the single-node search structure example (Figure 2.1) from Chapter 2. We could describe the state of the structure by the proposition

$$r \mapsto v,$$

which means r is a heap location containing value v (in a real proof v will be existentially quantified, but there will also be other propositions conjoined to ensure that it is not an arbitrary value). Such a description would be quite limiting: we would allow implementations only where each node contains a single heap location, implying that the search structure would become full as soon as a key is added to it, and we would not be able to provide implementations for lockNode and unlockNode as most lock implementations require additional memory (e.g., a bit to hold the lock state). To solve the second issue, we could instead use the proposition

$$r \mapsto v \, * \, \mathsf{lk}(r) \mapsto b,$$

where the second heap location (at address $\mathsf{lk}(r)$) stores the lock bit. Note that the use of the separating conjunction $*$ implies that $r \neq \mathsf{lk}(r)$, and so this proposition describes two distinct heap locations. If we wanted to describe an implementation where our single node contained an array of length $N + 1$, we could use the proposition

$$r \mapsto v_0 \, * \, (r+1) \mapsto v_1 \, * \, \ldots \, * \, (r+N) \mapsto v_N \, * \, \mathsf{lk}(r) \mapsto b,$$

which describes $N + 1$ consecutive heap locations and a distinct lock location. This can also be expressed succinctly using the iterated separating conjunction:

$$\mathsf{lk}(r) \mapsto b \, * \, \operatorname*{\text{\LARGE$*$}}_{0 \leq i \leq N} (r+i) \mapsto v_i$$

For the sake of generality, we want to leave the concrete details of how the node stores its keys to the implementation of the template algorithm. This suggests we need to use some kind of a template proposition, one that can be instantiated to capture the instantiations described above as well as others. This can be achieved by using an *abstract predicate* $\mathsf{Node}(n, C_n)$, which stands for an unspecified implementation of a node at address n containing keys C_n. Our template proof can then describe the state of the search structure by the proposition

$$\mathsf{Node}(r, C) \, * \, \mathsf{lk}(r) \mapsto b,$$

which describes a state containing a single node r, with contents C, and a lock flag at location $\mathsf{lk}(r)$ that has some contents b. Note that predicates express both ownership of a resource (in

$$\Psi_\omega(k, C, C', res) := \begin{cases} C' = C \wedge (res \iff k \in C) & \omega = \texttt{search} \\ C' = C \cup \{k\} \wedge (res \iff k \notin C) & \omega = \texttt{insert} \\ C' = C \setminus \{k\} \wedge (res \iff k \in C) & \omega = \texttt{delete} \end{cases}$$

Figure 3.3: The specification of core search structure operations. For core operation ω on key k, Ψ_ω describes the relation between C, the contents before the operation, C', the contents after the operation, and the return value res.

the case of $\mathsf{Node}(n, C_n)$, the node n) as well as data abstraction (C_n is the contents, or abstract value, of node n) [Hoare, 1972].

Our template proof then specifies the assumptions it makes about the predicate. As long as implementations of Node satisfy these assumptions, the template proof will be valid. For instance, all our template proofs assume that

$$\forall n, C_n, C_n'. \; \mathsf{Node}(n, C_n) \; * \; \mathsf{Node}(n, C_n') \; \twoheadrightarrow \mathsf{False},$$

an analog to the property of points-to predicates ($x \mapsto y \; * \; x \mapsto z \; \twoheadrightarrow \mathsf{False}$) that captures the intuition that the $\mathsf{Node}(n, C_n)$ predicate "owns" the heap address n, which cannot be in two places at once.

Single-Node Implementation Specification.   In the single-node template, we can use the abstract predicate Node to specify the behavior the template expects from the helper function `decisiveOp` using a Hoare triple:

$$\{\mathsf{Node}(r, C)\} \; \texttt{decisiveOp } \omega \; r \; k \; \{res. \, \mathsf{Node}(r, C') \; * \; \Psi_\omega(k, C, C', res)\}$$

The predicate Ψ_ω, defined in Figure 3.3, captures the behavior of the set abstract data type, and is the abstract specification of the single-copy search structures that we study in this monograph. For a given search structure operation ω on key k, $\Psi_\omega(k, C, C', res)$ relates the contents of the search structure before the operation (C), to the contents after the operation (C'), and the return value res. The Hoare triple for `decisiveOp` says: given a state containing a node r with contents C, the operation ω returns a value res and the node r has (new) contents C' such that Ψ_ω expresses the relationship between the old and new contents and res. For example, an insert operation adds k to the contents, and returns true if and only if k was not already present in the search structure.

Single-Node Implementation Proof.   For example, the simple implementation for the single-node template that we discussed before is one that uses a single heap cell for the node r. Figure 3.4 shows the definition of Node in this case, which uses a default value $\perp$ to indicate that

```
1 Node(n, C) := ∃v. n ↦ v  *  C = {v} \ {⊥}
2
3 {Node(r, C)}
4 let rec insert r k =
5    {∃v. r ↦ v  *  C = {v} \ {⊥}}
6    let u = !r in
7    {∃v. r ↦ v  *  C = {v} \ {⊥}  *  u = v}
8    if u == ⊥ then
9       {r ↦ ⊥  *  C = ∅}
10      r ← k;
11      {r ↦ k  *  C = ∅}
12      true
13      {Node(r, {k})  *  C = ∅}
14   else
15      {∃v. r ↦ v  *  C = {v} \ {⊥}}
16      {Node(r, C)}
17      insert r k
18      {v. Node(r, C')  *  Ψ_ω(k, C, C', v)}
19 {v. Node(r, C')  *  Ψ_ω(k, C, C', v)}
```

Figure 3.4: Proof of insert for a single-node implementation.

the heap cell is empty, and relates the value in the heap cell to the contents of the node if non-empty. The figure also shows a proof that the insert operation, which inserts the given value if the heap cell is empty and loops indefinitely otherwise,[6] satisfies the specification for decisive operations given in Figure 4.5.

3.5   INVARIANTS

Consider the following concurrent program:[7]

$$e_{\text{even}} := \textbf{let } x = \textbf{ref}(0) \textbf{ in } (x \leftarrow 2 \parallel x \leftarrow 4); \, !x$$

And say we want to prove the specification

$$\{\text{True}\} \, e_{\text{even}} \, \{v. \, \text{Even}(v)\},$$

where $\text{Even}(n) := (n \% 2 = 0)$ says that the program returns an even number. Unlike e_{par}, this program manipulates the *same* heap location in two parallel threads. This means we cannot use

---

[6]All proofs in this monograph are concerned with *partial* correctness, i.e., if a program terminates then it satisfies the given specification. Our methodology can be extended (with additional assumptions) to also prove termination, as we discuss in the future work section of the last chapter.

[7]This example is adapted from Chajed et al. [2021].

the parallel composition rule HOARE-PAR, as there is no way we can duplicate the heap cell $x \mapsto 0$ in order to give it to both of the threads:

$$
\begin{array}{c}
\{\mathsf{True}\} \\
\textbf{let } x = \textbf{ref}(0) \textbf{ in} \\
\{x \mapsto 0\} \\
(* \; \mathtt{IMPOSSIBLE} \; *) \\
\{x \mapsto 0 \; * \; x \mapsto 0\} \\
\begin{array}{c|c}
\{x \mapsto 0\} & \{x \mapsto 0\} \\
x \leftarrow 2 & x \leftarrow 4 \\
\{x \mapsto 2\} & \{x \mapsto 4\}
\end{array} \\
\{\ldots\} \\
!x \\
\{v. \ldots\}
\end{array}
$$

Nevertheless, the program is safe, and no matter which thread writes to x first, the returned value will be even. The program does not crash because the threads are using atomic instructions to store a value at x, so despite the fact that x is a shared heap cell, both stores are safe. Moreover, since both threads write an even number to x, the final result is also even. We need a proof mechanism that allows us to share state between threads and formalize the above reasoning.

The solution is to use invariants. An *invariant* in Iris is a proposition of the form $\boxed{P}^{N}$, where P is an arbitrary Iris proposition. Invariants provide a mechanism to reason about ownership of resources describing shared state that can be concurrently accessed by many threads. Intuitively, an invariant is a property that, once established, will remain true forever. It is therefore a duplicable resource and can be freely shared with any thread.

However, in order to ensure that the invariant indeed remains valid once it has been established, Iris' proof rules for invariants impose restrictions on how the resources contained in an invariant can be accessed and manipulated. At any point in time, a thread can *open* an invariant $\boxed{P}^{N}$ and gain ownership of the contained resources P. These resources can then be used in the proof of a single atomic step of the thread's execution. After the thread has performed an atomic step with an open invariant, the invariant must be *closed*, which amounts to proving that P has been reestablished. Otherwise, the proof cannot succeed.

The proof rules for invariants are given in Figure 3.5. The $\triangleright$ symbol is called the "later" modality, and is needed to ensure soundness in complex cases when Hoare triples or invariants themselves are stored inside invariants; we can ignore it for now. INV-ALLOC is a rule that allows us to take a resource R that we own and turn it into an invariant. In most proofs, this signifies the point at which some local state is shared with other threads for the first time. INV-DUP allows invariants to be freely duplicated; this will allow us to share them among threads. Finally, INV-

INV-ALLOC
$$\frac{\left\{\boxed{R}^{\mathcal{N}} * P\right\} e \left\{Q\right\}_{\mathcal{E}}}{\left\{\triangleright R * P\right\} e \left\{Q\right\}}$$

INV-DUP
$$\boxed{R}^{\mathcal{N}} \vdash \boxed{R}^{\mathcal{N}} * \boxed{R}^{\mathcal{N}}$$

INV-OPEN
$$\frac{\left\{\triangleright R * P\right\} e \left\{\triangleright R * Q\right\}_{\mathcal{E}} \qquad e \text{ atomic}}{\left\{\boxed{R}^{\mathcal{N}} * P\right\} e \left\{\boxed{R}^{\mathcal{N}} * Q\right\}_{\mathcal{E} \uplus \mathcal{N}}}$$

Figure 3.5: Proof rules for Iris invariants.

OPEN allows us to open an invariant and gain ownership of its contents for the duration of an atomic step.

The $\mathcal{N}$ in $\boxed{P}^{\mathcal{N}}$ refers to the *namespace* of the invariant. Namespaces are part of the mechanism used in Iris to keep track of invariants that are currently open and need to be closed before the next atomic step. This is necessary to avoid issues of re-entrancy in case of nested invariants, which would lead to logical inconsistencies. The set of namespaces that one is allowed to open is kept as an annotation (subscript) to the Hoare triple in question; for instance, note that INV-OPEN ensures that the invariant named $\mathcal{N}$ cannot be opened again. Since most proofs in this monograph use only a single invariant, we also omit these namespace annotations from Hoare triples.

We can now prove that e_{even} returns an even number, using an invariant (see Figure 3.6). Here we use INV-ALLOC at the beginning to transfer the newly created heap cell $x \mapsto 0$ into an invariant. We then use INV-DUP to duplicate it, and then share it among the two threads using HOARE-PAR. Each thread then uses INV-OPEN to open the invariant around the atomic store operation, after which we show that the store preserves the invariant that x contains an even number in each case. After the threads join, we once again open the invariant to read the contents of x, and the invariant tells us that whatever the value is, it is even.[8]

3.6 ATOMIC TRIPLES

In this section we introduce the concept of *logical atomicity*, which is used to specify programs that execute in multiple atomic steps but whose effect appears to take place in a single point in time. This concept will be useful for specifying concurrent search structures in a way that can be used to verify concurrent client programs.

To see why we need a new form of specification, suppose we tried to specify the behavior of a concurrent search structure using a Hoare triple as follows (recall from the single-node template algorithm in Figure 2.1 that cssOp is a function subsuming all the core search structure operations):

$$\{\text{CSS}(r, C)\} \text{ cssOp } \omega \ r \ k \ \{res. \ \text{CSS}(r, C') \ * \ \Psi_{\omega}(k, C, C', res)\}, \tag{3.1}$$

---

[8]We throw away the invariant at the end because we no longer need it, which we can do because Iris is affine.

$$\{\mathsf{True}\}$$
$$\mathbf{let}\ x\ =\ \mathbf{ref}(0)\ \mathbf{in}$$
$$\{x \mapsto 0\}$$
$$(*\ \texttt{Allocate invariant}\ *)$$
$$\boxed{\exists n.\ x \mapsto n\ *\ \mathsf{Even}(n)}^{\mathcal{N}}$$

$$\left\{ \boxed{\exists n.\ x \mapsto n\ *\ \mathsf{Even}(n)}^{\mathcal{N}}\ *\ \boxed{\exists n.\ x \mapsto n\ *\ \mathsf{Even}(n)}^{\mathcal{N}} \right\}$$

INV-OPEN	$\left\{\boxed{\exists n.\ x \mapsto n\ *\ \mathsf{Even}(n)}^{\mathcal{N}}\right\}$ $\{x \mapsto n\ *\ \mathsf{Even}(n)\}$ $x \leftarrow 2$ $\{x \mapsto 2\ *\ \mathsf{Even}(2)\}$	‖	INV-OPEN	$\left\{\boxed{\exists n.\ x \mapsto n\ *\ \mathsf{Even}(n)}^{\mathcal{N}}\right\}$ $\{x \mapsto n\ *\ \mathsf{Even}(n)\}$ $x \leftarrow 4$ $\{x \mapsto 4\ *\ \mathsf{Even}(4)\}$

$$\left\{\boxed{\exists n.\ x \mapsto n\ *\ \mathsf{Even}(n)}^{\mathcal{N}}\right\}\ \|\ \left\{\boxed{\exists n.\ x \mapsto n\ *\ \mathsf{Even}(n)}^{\mathcal{N}}\right\}$$

$$\left\{\boxed{\exists n.\ x \mapsto n\ *\ \mathsf{Even}(n)}^{\mathcal{N}}\ *\ \boxed{\exists n.\ x \mapsto n\ *\ \mathsf{Even}(n)}^{\mathcal{N}}\right\}$$

$$\left\{\boxed{\exists n.\ x \mapsto n\ *\ \mathsf{Even}(n)}^{\mathcal{N}}\right\}$$

INV-OPEN:
$$\{x \mapsto n\ *\ \mathsf{Even}(n)\}$$
$$!x$$
$$\{v.\ x \mapsto v\ *\ \mathsf{Even}(v)\}$$
$$\left\{v.\ \boxed{\exists n.\ x \mapsto n\ *\ \mathsf{Even}(n)}^{\mathcal{N}}\ *\ \mathsf{Even}(v)\right\}$$
$$\{v.\ \mathsf{Even}(v)\}$$

Figure 3.6: Invariant-based proof that e_{even} returns an even number.

where $\mathsf{CSS}(r, C)$ (for concurrent search structure) is a predicate describing a search structure at location r with contents C. While we might be able to prove that the single-node template satisfies this specification, this specification will not be helpful when reasoning about concurrent programs that use the single-node search structure.

To see why, consider the client program:

$$e_{\mathsf{client}} := (\mathsf{cssOp\ insert}\ r\ k_1\ \|\ \mathsf{cssOp\ insert}\ r\ k_2)$$

This is a program that manipulates a concurrent search structure rooted at r by calling the insert operation on two keys k_1 and k_2 on two parallel threads.

A simple specification that captures the fact that e_{client} is memory-safe in a concurrent context is:

$$\{\mathsf{CSS}(r, C)\}\ e_{\mathsf{client}}\ \{\mathsf{True}\}$$

One might try to prove this specification the same way we proved that e_{even} returned an even number, by creating a new invariant containing $\mathsf{CSS}(r, C)$ and sharing it among the two threads.

However, there is a catch: we cannot use INV-OPEN to open the invariant when we need to get $CSS(r, C)$ to satisfy the precondition of cssOp in (3.1). The reason is that INV-OPEN can be applied only to programs that are *physically atomic*, i.e., that they execute in a single machine instruction (e.g., $!x$, or a **CAS**). Here, on the other hand, we need to reason about cssOp, which for most realistic search structures executes in multiple physical steps.

On the other hand, a good concurrent search structure is designed to be used in precisely such conditions, by using locks or other concurrent protocols to ensure that concurrent invocations of cssOp from multiple threads is safe. We thus need a stronger specification for cssOp, one that captures the fact that in concurrent settings cssOp *behaves* as though it is atomic.

We can specify the concurrent behavior of such programs using *atomic triples* [da Rocha Pinto et al., 2014, Jung et al., 2020, 2015]. An atomic triple $\langle x.\ P \rangle\ e\ \langle v.\ Q \rangle$ is made up of the precondition P (that can refer to the *pseudo-quantified* variable x, as explained below), return value v, postcondition Q (that can refer to v and x), and a program e. Such a triple means that e, despite executing in potentially many atomic steps, appears to operate atomically on the shared state and transforms it from a state satisfying P to one satisfying Q.

Atomic triples are strongly related to the well-known *linearizability* [Filipovic et al., 2009, Herlihy and Wing, 1990] criterion for concurrent algorithms. Intuitively, there is a point in time during the course of the execution of e, known as the *linearization point*, where e updates P to Q. For the example of an insert operation on a search structure, this will be the point of time when the inserted value is visible to other threads. Linearizability requires that a concurrent set of operations produces the same final state and returns the same values as a sequential execution of the operations where the ordering is the order of the linearization points. In other literature [Bernstein et al., 1987], linearizability is known as order-preserving serializability.

The specification we want to prove for concurrent search structures is the following:

$$\langle C.\ CSS(r, C) \rangle\ \text{cssOp}\ \omega\ r\ k\ \langle res.\ CSS(r, C')\ *\ \Psi_\omega(k, C, C', res) \rangle \tag{3.2}$$

The binder on C in the precondition is a special *pseudo-quantifier* that captures the fact that during the execution of ω, the value of C can change (e.g., by concurrent operations). At the linearization point however, cssOp changes $CSS(r, C)$ to $CSS(r, C')$ in an atomic step. The new set of keys C' and the eventual return value res satisfy the predicate $\Psi_\omega(k, C, C', res)$. Note that the C in the postcondition is bound in the precondition, i.e., to the contents *just before* the linearization point. The goal is that clients of the search structure can pretend that they are using a serial or sequential implementation with specification Ψ_ω.

We call operations that satisfy atomic triples as being *logically atomic*. Coming back to our motivating example e_{client}, once we have proved that cssOp is logically atomic, we can use the following rule to open an invariant around it:

LOGATOM-INV

$$\frac{\langle R * P \rangle\ e\ \langle v.\ R * Q(v) \rangle}{\boxed{R}^N \vdash \langle P \rangle\ e\ \langle v.\ Q(v) \rangle}$$

This allows us to complete the proof of e_{client} using an invariant and (3.2):

$$\{\text{CSS}(r, C)\}$$
$$(*\ \texttt{Allocate invariant}\ *)$$
$$\left\{\boxed{\exists C.\ \text{CSS}(r, C)}^{\,\mathcal{N}}\right\}$$
$$\left\{\boxed{\exists C.\ \text{CSS}(r, C)}^{\,\mathcal{N}}\ *\ \boxed{\exists C.\ \text{CSS}(r, C)}^{\,\mathcal{N}}\right\}$$

$$\left\{\boxed{\exists C.\ \text{CSS}(r, C)}^{\,\mathcal{N}}\right\} \qquad\qquad \left\{\boxed{\exists C.\ \text{CSS}(r, C)}^{\,\mathcal{N}}\right\}$$

LOGATOM-INV $\langle\, \text{CSS}(r, C) \,\rangle$ $\qquad$ LOGATOM-INV $\langle\, \text{CSS}(r, C) \,\rangle$
$\texttt{cssOp insert } r\ k_1$ $\qquad\qquad$ $\texttt{cssOp insert } r\ k_2$
$\langle\, \text{CSS}(r, C \cup \{k_1\}) \,\rangle$ $\qquad$ $\langle\, \text{CSS}(r, C \cup \{k_2\}) \,\rangle$

$$\left\{\boxed{\exists C.\ \text{CSS}(r, C)}^{\,\mathcal{N}}\right\} \qquad\qquad \left\{\boxed{\exists C.\ \text{CSS}(r, C)}^{\,\mathcal{N}}\right\}$$
$$\left\{\boxed{\exists C.\ \text{CSS}(r, C)}^{\,\mathcal{N}}\ *\ \boxed{\exists C.\ \text{CSS}(r, C)}^{\,\mathcal{N}}\right\}$$
$$\{\text{True}\}$$

Note that the invariant that we used in the above proof was

$$\boxed{\exists C.\ \text{CSS}(r, C)}^{\,\mathcal{N}},$$

which existentially quantifies over the search structure contents C. This means that each time we open the invariant, the search structure can potentially have a different set of contents. This also means that when we close the invariant, we "forget" any changes we made to the contents, for example, that the left thread added k_1 to the contents. This weak invariant (which does not place any constraints on the contents) is sufficient because the postcondition here is simply True. If we wanted to prove something stronger, for example that the contents at the end of e_{client} will be $C \cup \{k_1, k_2\}$, then we would need a more complex invariant. In fact, we would need some way of keeping track of the updates the two threads make to the structure, for which we need to use ghost state (a concept we introduce in Chapter 4).

3.7   PROVING ATOMIC TRIPLES

Let us now turn to the question of how to prove that a program satisfies an atomic triple specification.

Recall that an atomic triple $\langle x.\ P \rangle\, e\, \langle v.\ Q \rangle$ means there is a single physical step during the execution of e when the shared state is transformed from P to Q. Thus, while proving $\langle x.\ P \rangle\, e\, \langle v.\ Q \rangle$, we cannot treat P and Q as the pre- and postconditions of e as a whole (remember, e is potentially a complex program consisting of multiple atomic steps). It is more accurate to think of P and Q as the pre- and postcondition to e's linearization point. Unlike proofs of Hoare triples, where we are given ownership to the resources in P at the beginning of e's execution and are under an obligation to transform them to Q by the end, in proofs of

$$\text{LOGATOM-INTRO} \atop \dfrac{\forall \Phi. \; \{\text{AU}_{x.P,Q}(\Phi)\} \; e \; \{v. \; \Phi(v)\}}{\langle x. \; P \rangle \; e \; \langle v. \; Q \rangle}$$

$$\text{LOGATOM-ATOM} \atop \dfrac{\forall x. \; \{P\} \; e \; \{v. \; Q\} \qquad e \text{ atomic}}{\langle x. \; P \rangle \; e \; \langle v. \; Q \rangle}$$

$$\text{AU-ABORT} \atop \dfrac{\langle x. \; P * P' \rangle \; e \; \langle v. \; P * Q' \rangle}{\{\text{AU}_{x.P,Q}(\Phi) * P'\} \; e \; \{v. \; \text{AU}_{x.P,Q}(\Phi) * Q'\}}$$

$$\text{AU-COMMIT} \atop \dfrac{\langle x. \; P * P' \rangle \; e \; \langle v. \; Q * Q' \rangle}{\{\text{AU}_{x.P,Q}(\Phi) * P'\} \; e \; \{v. \; \Phi(v) * Q'\}}$$

$$\text{LOGATOM-FRAME} \atop \dfrac{\langle x. \; P \rangle \; e \; \langle v. \; Q \rangle}{\langle x. \; P * R \rangle \; e \; \langle v. \; Q * R \rangle}$$

Figure 3.7: Proof rules for establishing atomic triples.

atomic triples we can read or modify the resources in the precondition only during atomic steps. Furthermore, our obligation is that all atomic steps accessing P either make a modification that preserves P, except for (exactly) one step, which has to transform it into Q.

Figure 3.7 contains the proof rules that help us execute the above proof argument. Most proofs of atomic triples start by using the rule LOGATOM-INTRO, which converts the atomic triple into a standard Hoare triple. The precondition contains an *atomic update token* $\text{AU}_{x.P,Q}(\Phi)$, which records the fact that we are proving an atomic triple with precondition $x. \; P$ (recall, x is bound by a pseudo-quantifier) and postcondition Q. As we will see, this token gives us the *right* to use the resources in the precondition P when executing atomic instructions, but the token also records our *obligation* to transform P to Q before execution completes. One way to use the resources in P is to use the AU-ABORT rule, which gives us access to the precondition P if the expression e is atomic, i.e., if we can prove that e atomically transforms a global (shared) precondition P and some local precondition P' into a local postcondition Q' while leaving P unchanged. This rule is useful when a program has some initial operations that modify the shared state in a way that does not change the abstract state (for instance, by locking or performing maintenance on a node). At some point, however, the program must update the shared state to the postcondition Q. LOGATOM-INTRO enforces this obligation by using an unknown, universally quantified proposition $\Phi(v)$, in the postcondition of the Hoare triple. One can think of $\Phi(v)$ as the precondition for the continuation of the computation performed after e terminates, by a larger program containing e. The only way to prove $\Phi(v)$ is to use rule AU-COMMIT, which lets us exchange the atomic update token for $\Phi(v)$ if we can prove that the program e transforms the global state from P to Q in an atomic step. As with AU-ABORT, the rule allows for some extra local resources P' and Q' in the pre- and postcondition of e.

let unlockNode x =
$\langle \mathsf{lk}(x) \mapsto true \rangle$
$\quad \{ \mathsf{AU}_{P,Q}(\Phi) \}$
$\qquad \langle \mathsf{lk}(x) \mapsto true \rangle$
$\qquad\quad \{ \mathsf{lk}(x) \mapsto true \}$
$\qquad\quad \mathsf{lk}(x) \leftarrow false$
$\qquad\quad \{ \mathsf{lk}(x) \mapsto false \}$
$\qquad \langle \mathsf{lk}(x) \mapsto false \rangle$
$\quad \{ \Phi \}$
$\langle \mathsf{lk}(x) \mapsto false \rangle$

(vertical brackets labeled LOGATOM-INTRO, AU-COMMIT, LOGATOM-ATOM)

Figure 3.8: Proof of the unlock method.

To illustrate proofs of atomic triples, we prove that the lock and unlock methods from Figure 2.1 satisfy the following atomic specifications:

$$\langle b.\ \mathsf{lk}(x) \mapsto b \rangle \ \texttt{lockNode}\ x\ \langle \mathsf{lk}(x) \mapsto true\ *\ b = false \rangle$$

$$\langle \mathsf{lk}(x) \mapsto true \rangle \ \texttt{unlockNode}\ x\ \langle \mathsf{lk}(x) \mapsto false \rangle$$

This captures the behavior that `lockNode` atomically sets $\mathsf{lk}(x)$ to *true*, but only if it used to be *false*, and `unlockNode` sets it back to *false* if it used to be *true*.

First, let us look at the simpler method, `unlockNode` (Figure 3.8). As mentioned above, we start by converting the atomic triple into a Hoare triple using LOGATOM-INTRO.[9] Since `unlockNode` has only a single operation, this must be the linearization point, and so we use AU-COMMIT to convert the $\mathsf{AU}_{P,Q}(\Phi)$ resource into the precondition $\mathsf{lk}(x) \mapsto true$. We call this proof step of using AU-ABORT or AU-COMMIT to access resources in the precondition as *opening the precondition*. We then use LOGATOM-ATOM to perform the update on heap address $\mathsf{lk}(x)$, which gives us the state expected by the postcondition of `unlockNode`. Thus, we can successfully finish the application of the AU-COMMIT rule, which we call *committing the change*. This gives us the expected Hoare postcondition Φ (argument omitted since it is the unit), which allows us to complete the application of LOGATOM-INTRO, completing the proof.

The proof of `lockNode` is a bit more complicated, so we represent it in the inline style (Figure 3.9). Once again, we start with LOGATOM-INTRO to convert the atomic triple to a Hoare triple. Since the first instruction in the program is a **CAS** on a location $\mathsf{lk}(x)$ in the precondition, we must open the precondition somehow. The tricky part is that we do not know whether to use AU-ABORT or AU-COMMIT because we do not know if the **CAS** will succeed until we look at the shared state in the precondition.

---

[9]Note that for brevity we write P for the entire precondition, including any pseudo-quantifier, in all subsequent proofs.

```
 1  ⟨b. lk(x) ↦ b⟩
 2    {AU_{P,Q}(Φ)}  (* LOGATOM-INTRO *)
 3    let rec lockNode x =
 4      {AU_{P,Q}(Φ)}
 5      ⟨b. lk(x) ↦ b⟩  (* AU-ABORT or AU-COMMIT *)
 6        {lk(x) ↦ b}  (* LOGATOM-ATOM *)
 7      if CAS(lk(x), false, true) then
 8          {lk(x) ↦ true * b = false}  (* HOARE-CAS-SUC *)
 9          ⟨lk(x) ↦ true * b = false⟩  (* end of LOGATOM-ATOM *)
10          {Φ}  (* end of AU-COMMIT *)
11          ()
12      else
13          {lk(x) ↦ b}  (* HOARE-CAS-FAIL *)
14          ⟨lk(x) ↦ b⟩  (* end of LOGATOM-ATOM *)
15          {AU_{P,Q}(Φ)}  (* end of AU-ABORT *)
16          lockNode x
17          {Φ}  (* Recursive call: by induction *)
18    {Φ}
19  ⟨lk(x) ↦ true * b = false⟩
```

Figure 3.9: Proof of the lock method.

We thus do a proof by cases: depending on the value of b (the current value of the lock location $lk(x)$), we apply the appropriate rule (this is shown as a single proof in Figure 3.9 for brevity). In the case where b is *false* (i.e., the lock is currently unlocked), the **CAS** will succeed and set $lk(x)$ to *true*, thereby locking the node. So we use AU-COMMIT, LOGATOM-ATOM, and HOARE-CAS-SUC to obtain the state in line 8. We can then commit the change and obtain $Φ$ as in the previous proof. On the other hand, if b is *true*, then we use AU-ABORT, LOGATOM-ATOM, and HOARE-CAS-FAIL. In this case, we have not modified the shared state, so we *close* the precondition using AU-ABORT and get back the $AU_{P,Q}(Φ)$ resource. We then deal with the recursive call to lockNode by induction, using the specification that we are trying to prove,

$$\{AU_{P,Q}(Φ)\}\ \text{lockNode}\ x\ \{Φ\},$$

as the induction hypothesis to complete this branch of the proof.

CHAPTER 4

Ghost State

Ghost state, originally called auxiliary variables [Owicki and Gries, 1976], is a formal technique where the prover adds state (variables or resources) to a program that capture knowledge about the history of a computation not present in the state of the original program in order to verify it. As long as the added ghost state, and the ghost commands that modify it, have no effect on the run-time behavior of the program, then a so-called *erasure* theorem states that a proof of the augmented program can be transformed into a proof of the program with all ghost state removed (i.e., the original program). In Iris, ghost state is purely logical and ghost commands are represented as proof rules, which gives us such an erasure theorem for free. The ghost state concept has shown itself to be an invaluable tool in the verifier's toolbox, and has been used to encode many common reasoning techniques including permissions, tokens, capabilities, and protocols.

In this chapter, we explain the technique of using ghost state to construct proofs of concurrent algorithms and describe how we use ghost state in Iris to verify template search structures. We start by motivating the need for ghost state using the single-node template algorithm as an example. We then see how Iris supports ghost state via the notion of resource algebras (RAs), and use a fractional RA to verify the single-node template.

Our treatment of ghost state in Iris is restricted to RAs that have units, as that is sufficient for the proofs in this monograph. Iris supports more general and powerful kinds of ghost state, so-called *cameras*, and we refer interested readers to the introduction to cameras by Jung et al. [2018]. We also motivate ghost state in this chapter as a means to verifying template search structures, but ghost state has a wide variety of applications in verifying many kinds of algorithms and properties. The Iris tutorial at POPL 2021 [Chajed et al., 2021] contains motivating examples that demonstrate other uses of ghost state.

4.1   MOTIVATION

The single-node template algorithm (Chapter 2, Figure 2.1) performs a core search structure operation on a structure consisting of a single node by using a lock to prevent unwanted interference from concurrent operations. Our task is to prove that the single-node template satisfies the specification:

$$\langle C.\ \mathsf{CSS}(r, C) \rangle\ \mathtt{cssOp}\ \omega\ r\ k\ \langle res.\ \mathsf{CSS}(r, C')\ *\ \Psi_\omega(k, C, C', res) \rangle$$

We can use the following specification for the `lockNode` and `unlockNode` (proved in Chapter 3):

$$\langle\, b.\ \mathsf{lk}(x) \mapsto b \,\rangle\ \texttt{lockNode}\ x\ \langle\, \mathsf{lk}(x) \mapsto \mathit{true}\ *\ b = \mathit{false} \,\rangle$$

$$\langle\, \mathsf{lk}(x) \mapsto \mathit{true} \,\rangle\ \texttt{unlockNode}\ x\ \langle\, \mathsf{lk}(x) \mapsto \mathit{false} \,\rangle$$

In this single-node protocol, all threads modify the node only when they have locked the node. This ensures that despite the fact that `decisiveOp` may consist of multiple instructions and therefore may not be atomic, its effect will appear to be atomic to other threads. We need a way to encode this protocol into the definition of CSS.

In §3.4, we proposed using an abstract predicate $\mathsf{Node}(n, C_n)$ to represent the single node in the template proof, because the structure of the node (whether it consists of one heap location, or many, etc.) is implementation-specific. Based on this, a first attempt at a definition for CSS would be:[1]

$$\mathsf{CSS}_1(r, C) := \exists b.\ \mathsf{lk}(r) \mapsto b\ *\ (b\ ?\ \mathsf{True} : \mathsf{Node}(r, C))$$

Here, the second term is $\mathsf{Node}(r, C)$ if b is *false* (the node is unlocked), indicating that unlocked nodes belong to the shared state and are available for anyone to acquire. But if b is *true* (the node is locked), then the shared state contains only True, which means that the thread that locked the node can take possession of the $\mathsf{Node}(r, C)$ resource into its local state.

Figure 4.1 presents a proof attempt that shows this locking mechanism in action. We start by using LOGATOM-INTRO, which converts our goal to the corresponding Hoare triple (we use P and Q to denote the pre- and postcondition of `cssOp` for brevity). The first line of code in `cssOp` is the call to `lockNode`, whose specification tells us that it needs the resource $\mathsf{lk}(r) \mapsto \_$. As this heap cell is inside CSS_1, which is inside the precondition P, we use AU-ABORT to get access to P. Note that we do not use AU-COMMIT, since this is not the linearization point of this method. We then use LOGATOM-FRAME to frame $\mathsf{Node}(r, C)$, which is not modified by `lockNode`, and reduce the proof to exactly the specification of `lockNode`.

When we finish the proof of the Hoare triple in the premise of rule AU-ABORT, the proof context is $\mathsf{CSS}_1(r, C)\ *\ \mathsf{Node}(r, C)$. Note that we have here instantiated AU-ABORT with $P = \mathsf{CSS}_1(r, C)$ and $Q' = \mathsf{Node}(r, C)$. Since we need to "give back" ownership of P only in AU-ABORT, our proof context after applying rule AU-ABORT contains a $\mathsf{Node}(r, C)$ predicate. We view any resources that are in an intermediate state with curly braces (such as $\mathsf{Node}(r, C)$), as being in the local state of the current thread. Non-atomic operations on local state (such as $\mathsf{Node}(r, C)$) can be made without risk of interference (i.e., without the risk that some other thread will read or modify that local state).

We are now back to proving a Hoare triple and can apply the Hoare triple specification of `decisiveOp` on $\mathsf{Node}(r, C)$ (after using HOARE-FRAME to frame the extra resource $\mathsf{AU}_{P,Q}(\Phi)$).

---

[1]This definition is closely related to the standard monitor invariants [Hoare, 1974], and their adaptation to separation logic and related formalisms is well known [Leino and Müller, 2009].

$CSS_1(r, C) := \exists b.\ \mathsf{lk}(r) \mapsto b \ * \ (b\ ?\ \mathsf{True} : \mathsf{Node}(r, C))$

$\langle C.\ CSS_1(r, C) \rangle$
let $\mathsf{cssOp}\ \omega\ r\ k\ =$

$\{AU_{P,Q}(\Phi)\}$

$\langle CSS_1(r, C) \rangle$

$\langle \exists b.\ \mathsf{lk}(r) \mapsto b \ * \ (b\ ?\ \mathsf{True} : \mathsf{Node}(r, C)) \rangle$ (* By definition *)

$\langle \exists b.\ \mathsf{lk}(r) \mapsto b \rangle$

$\mathsf{lockNode}\ r;$

$\langle \mathsf{lk}(r) \mapsto true \ * \ b = false \rangle$

$\langle \mathsf{lk}(r) \mapsto true \ * \ \mathsf{Node}(r, C) \rangle$

$\langle CSS_1(r, C) \ * \ \mathsf{Node}(r, C) \rangle$

$\{AU_{P,Q}(\Phi) \ * \ \mathsf{Node}(r, C)\}$

$\{\mathsf{Node}(r, C)\}$

let $res = \mathsf{decisiveOp}\ \omega\ r\ k$ **in**

$\{\mathsf{Node}(r, C') \ * \ \Psi_\omega(k, C, C', res)\}$

$\{AU_{P,Q}(\Phi) \ * \ \mathsf{Node}(r, C') \ * \ \Psi_\omega(k, C, C', res)\}$

$\langle CSS_1(r, C'') \ * \ \mathsf{Node}(r, C') \ * \ \Psi_\omega(k, C, C', res) \rangle$

$\langle \exists b.\ \mathsf{lk}(r) \mapsto b \ * \ (b\ ?\ \mathsf{True} : \mathsf{Node}(r, C'')) \ * \ \mathsf{Node}(r, C') \ * \ \Psi_\omega(k, C, C', res) \rangle$

$\langle \mathsf{lk}(r) \mapsto true \ * \ \mathsf{Node}(r, C') \ * \ \Psi_\omega(k, C, C', res) \rangle$ (* Property of Node *)

$\langle \mathsf{lk}(r) \mapsto true \rangle$

$\mathsf{unlockNode}\ r;$

$\langle \mathsf{lk}(r) \mapsto false \rangle$

$\langle \mathsf{lk}(r) \mapsto false \ * \ \mathsf{Node}(r, C') \ * \ \Psi_\omega(k, C, C', res) \rangle$

$\langle CSS_1(r, C') \ * \ \Psi_\omega(k, C, C', res) \rangle$

(* Problem: $C \neq C''$ *)

$\langle CSS_1(r, C') \ * \ \Psi_\omega(k, C'', C', res) \rangle$

$\{\Phi(res)\}$

res

$\{v.\ \Phi(v)\}$

$\langle v.\ CSS_1(r, C') \ * \ \Psi_\omega(k, C, C', v) \rangle$

(Left margin labels: LOGATOM-INTRO; AU-ABORT; LOGATOM-FRAME; HOARE-FRAME; AU-COMMIT; LOGATOM-FRAME)

Figure 4.1: First attempt at a proof of the single-node template.

We now turn our attention to the call to unlockNode. Observe that unlockNode modifies the lock location of r, whose heap cell is still part of the shared state (as it is contained in the definition of CSS_1), so we need to open the precondition again. But this step in the algorithm is also its linearization point, because when the thread unlocks the node that it has (potentially) modified, any changes it has made become visible to other threads.[2] We thus use the rule AU-COMMIT to open the precondition, which brings us to a subtle problem: recall that the precondition is $\langle C.\ \text{CSS}_1(r, C) \rangle$, where C is (pseudo) quantified. This means that every time we open the precondition, we get a potentially different set of contents C. Since we already have a variable called C in our context, when opening the precondition we get $\text{CSS}_1(r, C'')$ using a fresh variable C'' after using AU-COMMIT. The reason we get a fresh variable C'' is that $\text{CSS}_1(r, C)$ is in the *shared* state, and theoretically other threads can modify the contents of the search structure between the last time we opened the precondition (at the call to `lockNode`) and now. Although this cannot happen in our single-node example, where the current thread has locked the only node in the structure, the proof rules are generic rules that must apply to all algorithms, including search structures consisting of multiple nodes.

Unfortunately, AU-COMMIT needs us to show that `unlockNode` results in the postcondition of `cssOp` where the predicate $\Psi_\omega(k, C'', C', res)$ uses C'', the same variable we got when we opened the precondition. However, we have $\Psi_\omega(k, C, C', res)$ from the postcondition of `decisiveOp`, which uses C, the contents of the shared state during the call to `lockNode`. Thus, there is no way we can prove $\Psi_\omega(k, C'', C', res)$.

The real problem is in the definition of CSS_1:

$$\text{CSS}_1(r, C) := \exists b.\ \text{lk}(r) \mapsto b\ *\ (b\ ?\ \text{True} : \text{Node}(r, C))$$

Note that when the single node it contains is locked, this formula becomes

$$\text{lk}(r) \mapsto true\ *\ \text{True},$$

which says nothing about the contents C. This is why when we access the shared state for the second time around the call to `unlockNode`, we know nothing about the current contents C''.

In other words, the definition of CSS_1 encodes the protocol that threads take exclusive ownership of the node when locking it, but not the (perhaps obvious) knowledge that while a thread holds an exclusive lock on the node, no other thread can change the state of the search structure. Encoding such knowledge into the proof can be done by using ghost state, as explained below.

4.2   GHOST STATES AND RESOURCE ALGEBRAS

Iris expresses ownership of ghost state by the proposition $\ulcorner a \urcorner^\gamma$ which asserts ownership of a piece a of the ghost location γ. The values a stored in ghost state belong to a resource algebra (RA, defined formally below). For example, a *fractional* RA over sets of keys consists of elements of

---

[2]In fact, the linearization point will be at the point of unlocking for all the lock-based single-copy search structures that we study in this monograph. In Chapter 12, we will study a template where this is not the case.

the form (q, C), where $q \in (0, 1]$ is a rational number and C is a set of keys, as well as a unit element ε and a special element $\frac{\ell}{}$. As we will see, a fractional ghost location can be split and shared among many threads to permit shared reads, but threads can only write to the ghost location if they own the full ghost location. This means that they have the following properties:

$$\boxed{(q, C)}^{\gamma} * q = q_1 + q_2 \implies \boxed{(q_1, C)}^{\gamma} * \boxed{(q_2, C)}^{\gamma} \tag{4.1}$$

$$\boxed{(q_1, C_1)}^{\gamma} * \boxed{(q_2, C_2)}^{\gamma} \vdash C_1 = C_2 \tag{4.2}$$

Here, the first property tells us that if we own $\boxed{(q, C)}^{\gamma}$ then we can split it into smaller fractions (via a ghost update $\implies$, explained below). The ghost update implicitly assumes $q_1, q_2 \in (0, 1]$. The second property tells us that any two fractional states must agree on the set of keys.

Recall that to verify the single-node template, we need a way to keep track of the contents of the search structure in two places—the shared state and in the local state of the thread that locks the node—in such a way that the two "views" of the contents are the same. We can do this by creating a fractional ghost location and splitting it into two halves (using (4.1)), extending our definition of the search structure predicate to

$$\mathsf{CSS}(r, C) := \exists b. \boxed{\tfrac{1}{2}\, C}^{\gamma} * \mathsf{lk}(r) \mapsto b * \left(b \,?\, \mathsf{True} : \mathsf{Node}(r, C) * \boxed{\tfrac{1}{2}\, C}^{\gamma} \right),$$

where we write $\boxed{\tfrac{1}{2}\, C}^{\gamma}$ instead of $\boxed{(\tfrac{1}{2}, C)}^{\gamma}$ for brevity. Note that the choice of $\tfrac{1}{2}$ is arbitrary. The construction works for any split $\boxed{q\, C}^{\gamma}$ and $\boxed{(1 - q)\, C}^{\gamma}$. What is important is that while a thread owns one of the two parts of the ghost location γ, no other thread can change the contents C. That is, when a thread locks r, we give it both $\mathsf{Node}(r, C)$ and one of the halves $\boxed{\tfrac{1}{2}\, C}^{\gamma}$. When the thread looks at the shared state later, say when unlocking r, as explained in §4.1 the search structure will have a potentially different contents C''. However, with the added ghost state, the proof context will contain:

$$\boxed{\tfrac{1}{2}\, C''}^{\gamma} * \mathsf{lk}(r) \mapsto b * \ldots * \mathsf{Node}(r, C) * \boxed{\tfrac{1}{2}\, C}^{\gamma}$$

And using property (4.2), the proof can use the two halves $\boxed{\tfrac{1}{2}\, C''}^{\gamma} * \boxed{\tfrac{1}{2}\, C}^{\gamma}$ to conclude $C'' = C$.

Formally, a resource algebra (RA) consists of a set M, a *validity* predicate $\mathcal{V}(-)$, and a binary operation $(\cdot): M \times M \to M$ that satisfy the axioms in Figure 4.2 (*Prop* is the type of propositions of the meta-logic (e.g., Coq)). RAs are a generalization of the partial commutative monoid (PCM) algebra commonly used by separation logics.[3]

---

[3]Iris actually uses *cameras* as the structure underlying resources, but as we do not use higher-order resources (i.e., state which can embed propositions) in this monograph, we restrict our attention to resource algebras, a stronger, but simpler, structure. Furthermore, RAs technically have a *core* function $|-|$ that maps each element to a (potentially different) unit, but since all the RAs we use have a gobal unit (i.e., $|\_| = \varepsilon$), we omit the core function in our presentation and restrict our attention to unital RAs.

A *resource algebra* is a tuple $(M, \mathcal{V}: M \to Prop, (\cdot): M \times M \to M, \varepsilon \in M)$ satisfying:

$$\forall a, b, c. \ (a \cdot b) \cdot c = a \cdot (b \cdot c) \qquad \text{(RA-ASSOC)}$$
$$\forall a, b. \ a \cdot b = b \cdot a \qquad \text{(RA-COMM)}$$
$$\forall a. \ \varepsilon \cdot a = a \qquad \text{(RA-ID)}$$
$$\mathcal{V}(\varepsilon) \qquad \text{(RA-VALID-ID)}$$
$$\forall a, b. \ \mathcal{V}(a \cdot b) \Rightarrow \mathcal{V}(a) \qquad \text{(RA-VALID-OP)}$$

Figure 4.2: The definition of a (unital) resource algebra (RA).

The two important mechanisms for using ghost state are: (1) a ghost location can be split and combined using the rule: $\ulcorner a \urcorner^\gamma * \ulcorner b \urcorner^\gamma \dashv\vdash \ulcorner a \cdot b \urcorner^\gamma$; and (2) at any point, if a thread owns a resource $\ulcorner c \urcorner^\gamma$, then the value c is valid, i.e., $\mathcal{V}(c)$. This means that an RA's composition operator $\cdot$ must be associative and commutative.[4] The axiom RA-VALID-OP disallows taking an invalid element and composing it with another element to make it valid; since Iris maintains an invariant that the composition of all values in a ghost location is valid, this axiom implies that any sub-resource in that location is also valid. RA-ID makes ε an identity or unit element with respect to composition, and RA-VALID-ID says the unit must be valid.

One can think of the ghost state proposition $\ulcorner a \urcorner^\gamma$ as the ghost analog of the points-to predicate $x \mapsto v$ that asserts that the (real) location x contains value v.[5] However, $\ulcorner a \urcorner^\gamma$ asserts only that γ contains a value *one of whose parts* is a (as we saw with the fractional ghost state example above). This means ghost state can be split and combined according to the composition operator of the underlying RA.

Example 4.1   Given a set S, we define the fractional RA over values of S as:

$$M := (q, s) \in (\mathbb{Q} \cap (0, 1]) \times S \mid \varepsilon \mid \lightning \qquad \mathcal{V}(a) := a \neq \lightning \qquad \varepsilon \cdot a := a \cdot \varepsilon := a$$

$$(q_1, s_1) \cdot (q_2, s_2) := \begin{cases} (q_1 + q_2, s_1) & \text{if } q_1 + q_2 \leq 1 \wedge s_1 = s_2 \\ \lightning & \text{otherwise} \end{cases} \qquad \lightning \cdot \_ := \_ \cdot \lightning := \lightning$$

The fractional RA we used above in the single-node template proof sketch can be obtained by instantiating the RA in Example 4.1 with the powerset of the set of all keys. The definition of composition and validity in the fractional RA ensure that elements must agree on their second

---

[4]Readers familiar with separation algebras will notice that the composition operator is not partial; cases where composition used to be undefined can be encoded by sending them to an invalid element.

[5]In fact, Iris' core logic makes no distinction between ghost state and non-ghost state—the heap is represented using special ghost state, and the points-to predicate $x \mapsto v$ is defined using the ghost location predicate $\ulcorner a \urcorner^\gamma$. However, we continue to use the $\mapsto$ symbol for expressing constraints on heap locations as it is widely-used and will be familiar to readers who have studied SL before.

$$\frac{\text{GHOST-ALLOC}}{\mathcal{V}(a)}$$
$$\text{True} \Rrightarrow \exists \gamma. \ulcorner a \urcorner^{\gamma}$$

$$\text{GHOST-OP}$$
$$\ulcorner a \cdot b \urcorner^{\gamma} \Leftrightarrow \ulcorner a \urcorner^{\gamma} * \ulcorner b \urcorner^{\gamma}$$

$$\text{GHOST-VALID}$$
$$\ulcorner a \urcorner^{\gamma} \Rightarrow \mathcal{V}(a)$$

$$\frac{\text{GHOST-UPDATE}}{a \rightsquigarrow B}$$
$$\ulcorner a \urcorner^{\gamma} \Rrightarrow \exists b \in B. \ulcorner b \urcorner^{\gamma}$$

$$\frac{\text{VS-TRANS} \qquad P \Rrightarrow Q \qquad Q \Rrightarrow R}{P \Rrightarrow R}$$

$$\frac{\text{VS-FRAME} \qquad P \Rrightarrow Q}{P * R \Rrightarrow Q * R}$$

Figure 4.3: Proof rules for manipulating ghost resources and view shifts.

element and have compatible fractions. This implies that ghost locations containing elements of the fractional RA enjoy the properties (4.1) and (4.2).

Figure 4.3 provides proof rules for manipulating ghost resources. The rule GHOST-ALLOC is used to allocate a new ghost resource. The rule restricts the allocation to only those resources that are valid. We previously alluded to how rules GHOST-OP and GHOST-VALID can be useful when proving the single-node template. The remaining rules dictate how a ghost resource can be updated using the view shift modality $\Rrightarrow$, which we explain next through the example of the fractional RA.

Ghost state by itself is not very useful unless it can be updated. However, unlike physical state, which can be modified at any point to any value, ghost state updates are restricted since Iris maintains the invariant that the composition of all the pieces of ghost state at a particular location is valid (as given by $\mathcal{V}$). Iris allows only *frame-preserving updates $a \rightsquigarrow b$*, defined below.[6]

Definition 4.2   A *frame-preserving update* is a relation between an element $a \in M$ and a set $B \subseteq M$, written $a \rightsquigarrow B$, such that

$$\forall a_{\mathrm{f}} \in M. \ \mathcal{V}(a \cdot a_{\mathrm{f}}) \Rightarrow \exists b \in B. \ \mathcal{V}(b \cdot a_{\mathrm{f}}).$$

We write $a \rightsquigarrow b$ if $a \rightsquigarrow \{b\}$.

Intuitively, $a \rightsquigarrow b$ says that every *frame a_{f}* that is compatible with a should also be compatible with b. Thus, changing a thread's fragment of the ghost state from a to some b will not invalidate assumptions about a_{f} made by any other thread. The fractional RA has the following frame-preserving update:

$$\text{FRAC-UPD}$$
$$(1, s) \rightsquigarrow (1, s')$$

Note that the element $(1, s)$ has no frame (no non-unit element can compose with it), thus the frame-preserving update condition holds trivially.

[6]While most of the frame-preserving updates that we use result in a single element ($a \rightsquigarrow b$), we adopt the more general definition that takes an element to a set of possible new values ($a \rightsquigarrow B$) because it is used in §8.1.2.

This allows us to change the value stored at a ghost location holding a fractional RA value as long as we own all the pieces of that location. Correspondingly, we also note that there are no frame-preserving updates from (q, s) to any (q, s') when $q < 1$, which means no thread can change the value s unless that thread holds all the fragments (i.e., $q = 1$).

The rule GHOST-UPDATE in Figure 4.3 lifts frame-preserving updates on RA elements to ghost updates on Iris propositions, which are captured by Iris' view shift modality $\Rrightarrow$. The intuitive meaning of $P \Rrightarrow Q$ is that if we have the resource P, then we can perform one or more frame-preserving updates in order to transform the resource P to Q. Alternatively, we can regard $P \Rrightarrow Q$ as a Hoare triple, with P as precondition and Q as postcondition, but no program code, as P must be transformed to Q solely by manipulating ghost resources. The rule VS-TRANS enables combining multiple view shifts into one, while VS-FRAME allows *framing* out resources from the update, similar to Hoare triples. The rule HOARE-CSQ (introduced in Chapter 3) allows one to use view shifts to perform ghost updates on the pre- and postcondition of a Hoare triple.

4.3   PROOF OF THE SINGLE-NODE TEMPLATE

We now have all the tools we need to prove the single-node template (Figure 4.4).

For brevity, this and all following proof outlines of atomic triples omit mentioning usage of the rules LOGATOM-INTRO, AU-ABORT, and AU-COMMIT. All our proofs follow the same pattern: they use LOGATOM-INTRO to turn the target atomic triple into a Hoare triple, and then use AU-ABORT or AU-COMMIT for each operation in the program. We also omit the $AU_{P,Q}(\Phi)$ since it appears on every line, and instead just assume we have the resources in the precondition when we go from a Hoare triple to an atomic triple, like on line 7. We also use an additional predicate $N(n, C)$ for the combination of $Node(n, C)$ and the ghost state associated with that node.

Recall that our definition of CSS now uses fractional ghost state to keep track of the contents of the search structure:

$$\mathsf{CSS}(r, C) := \exists b.\; \boxed{½\,C}^{\gamma} \;*\; \mathsf{lk}(r) \mapsto b \;*\; \left(b \;?\; \mathsf{True} : \mathsf{Node}(r, C) \;*\; \boxed{½\,C}^{\gamma}\right).$$

We can picture the shared state as follows:

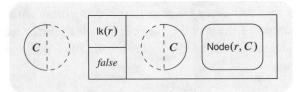

Here, we depict the two fractional ghost state elements as two complementary halves of a circle. We also depict the locked region by a rectangle containing, anti-clockwise from top-left, the address of the lock bit, the value of the lock bit, and the resources protected by the lock.

Consider the proof outline in Figure 4.4. The call to `lockNode` is handled as before, except that this time we take out $N(r, C)$ (i.e., both the node as well as its associated half of the contents)

```
1  N(n, C) := ⌜½ C⌝^γ  *  Node(n, C)

2  CSS(r, C) := ⌜½ C⌝^γ  *  ∃b. lk(n) ↦ b  *  (b ? True : N(r, C))

3

4  ⟨C. CSS(r, C)⟩

5  let cssOp ω r k =

6      {True}

7      ⟨⌜½ C⌝^γ  *  ∃b. lk(n) ↦ b  *  (b ? True : N(r, C))⟩

8      lockNode r;

9      ⟨⌜½ C⌝^γ  *  lk(n) ↦ true  *  N(r, C)⟩

10     {N(r, C)}

11     {Node(r, C)  *  ⌜½ C⌝^γ}

12     let res = decisiveOp ω r k in

13     {Node(r, C′)  *  ⌜½ C⌝^γ  *  Ψ_ω(k, C, C′, res)}

14     ⟨⌜½ C″⌝^γ  *  ∃b. lk(n) ↦ b  *  (b ? True : N(r, C″))  *  N(r, C′)  *  Ψ_ω(k, C, C′, res)⟩

15     ⟨⌜½ C″⌝^γ  *  lk(n) ↦ true  *  Node(r, C′)  *  ⌜½ C⌝^γ  *  Ψ_ω(k, C, C′, res)⟩

16     ⟨⌜½ C⌝^γ  *  lk(n) ↦ true  *  Node(r, C′)  *  ⌜½ C⌝^γ  *  Ψ_ω(k, C, C′, res)⟩

17     ⟨⌜½ C′⌝^γ  *  lk(n) ↦ true  *  Node(r, C′)  *  ⌜½ C′⌝^γ  *  Ψ_ω(k, C, C′, res)⟩

18     unlockNode r;

19     ⟨⌜½ C′⌝^γ  *  lk(n) ↦ false  *  Node(r, C′)  *  ⌜½ C′⌝^γ  *  Ψ_ω(k, C, C′, res)⟩

20     ⟨CSS(r, C′)  *  Ψ_ω(k, C, C′, res)⟩

21     {True}

22     res

23 ⟨res. CSS(r, C′)  *  Ψ_ω(k, C, C′, res)⟩
```

Figure 4.4: Proof of the single-node template algorithm.

from the precondition after locking r. The resulting proof context can be visualized as follows (where we denote the thread's local state in blue on the right):

The call to decisiveOp is also handled as before, and note that it modifies only the contents in Node(r, $C′$), the ghost state $⌜½\ C⌝^γ$ continues to record the contents of the shared state from the time when the node was locked. When we open the precondition around the call to `unlockNode`, we get the state shown on line 14, which we can visualize as:

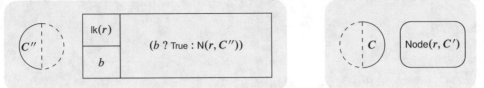

Note that the shared state uses a fresh variable C'' for the contents, and the lock bit's value is an unknown Boolean b. As before, we first use the property of the abstract predicate Node that $\mathsf{Node}(n, C_n) * \mathsf{Node}(n, C'_n) \twoheadrightarrow \mathsf{False}$ to infer that b must be *true* (line 15).

Crucially, we can now use the fractional resource algebra properties on the ghost state the thread owns at this point, $\lceil \frac{1}{2}\, C'' \rceil^\gamma * \lceil \frac{1}{2}\, C \rceil^\gamma$, to infer that $C = C''$. Intuitively, since the current thread had kept with it one half of the circle $\lceil \frac{1}{2}\, C \rceil^\gamma$, it knows that no one else could have changed the value on the other half $\lceil \frac{1}{2}\, C'' \rceil^\gamma$. The current proof context (line 16) thus looks like:

Before we unlock the node, we perform a frame-preserving update using FRAC-UPD to convert the ghost state to $\lceil \frac{1}{2}\, C' \rceil^\gamma * \lceil \frac{1}{2}\, C' \rceil^\gamma$, reflecting the updated contents of the search structure. We can then use the atomic specification of `unlockNode` to get the state shown on line 19:

We then use the specification of `unlockNode` to put back the node and its ghost state into the shared state, and show that we have the desired postcondition $\mathsf{CSS}(r, C') * \Psi_\omega(k, C, C', res)$:

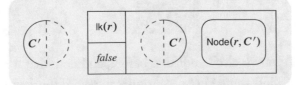

$$\{\mathsf{Node}(r,C)\}\ \mathtt{decisiveOp}\ \omega\ r\ k\ \{v.\,\mathsf{Node}(r,C')\ *\ \Psi_\omega(k,C,C',v)\}$$

$$\mathsf{Node}(n,C_n)\ *\ \mathsf{Node}(n,C_n')\ -\!\!*\ \mathsf{False}$$

Figure 4.5: The assumptions made by the single-node template on implementations.

$$\langle\, b\ R.\ \mathsf{L}(b,x,R)\,\rangle\ \mathtt{lockNode}\ x\ \langle\,\mathsf{L}(\mathit{true},x,R)\ *\ R\,\rangle$$

$$\langle\, R.\ \mathsf{L}(\mathit{true},x,R)\ *\ R\,\rangle\ \mathtt{unlockNode}\ x\ \langle\,\mathsf{L}(\mathit{false},x,R)\,\rangle$$

Figure 4.6: Abstract specification for `lockNode` and `unlockNode`.

The assumptions made by this template proof on implementation-specific helper functions and abstract predicates are listed in Figure 4.5. These functions and predicates are defined by implementations, which must satisfy the given assumptions. For example, we saw in §3.4 a proof that the simple implementation for the single-node template that uses a single heap cell for the node r satisfies the decisive operation specification.

Next, we will see how to extend these ideas to a structure consisting of more than one node.

4.4   ASIDE: ABSTRACT LOCK SPECIFICATION

Before we move on to template algorithms for multi-node structures, we take a brief aside and present an abstract specification for the `lockNode` and `unlockNode` methods so that we can reuse the part of the proof dealing with the locking mechanism.

We define an abstract higher-order predicate $\mathsf{L}(b,x,R)$ that captures a *lock region*:

$$\mathsf{L}(b,x,R) := \mathsf{lk}(x) \mapsto b\ *\ (b\ ?\ \mathsf{True}:R)$$

Here, R is a proposition that denotes an arbitrary resource protected by the lock with lock location $\mathsf{lk}(x)$ and lock bit b. The Boolean b indicates whether the lock is (un)locked. In terms of our visualizations from §4.3, we can visualize the predicate $\mathsf{L}(b,x,R)$ as:

We can use the lock region predicate to give higher-level specifications for `lockNode` and `unlockNode`, as shown in Figure 4.6. The specification of `lockNode` takes a lock region that

```
 1 ⟨b R. L(b, x, R)⟩
 2   let rec lockNode x =
 3     {True}
 4       ⟨b R. L(b, x, R)⟩ (* Open precondition *)
 5       {lk(x) ↦ b * (b ? True : R)} (* By definition *)
 6       if CAS(lk(x), false, true) then
 7         {lk(x) ↦ true * b = false * R}
 8         ⟨lk(x) ↦ true * b = false * R⟩
 9         ⟨L(true, x, R) * R⟩ (* Commit *)
10         {True}
11         ()
12       else
13         {lk(x) ↦ b * (b ? True : R)}
14         ⟨lk(x) ↦ b * (b ? True : R)⟩ (* Abort *)
15         {True}
16           ⟨b R. L(b, x, R)⟩ (* Open precondition *)
17           lockNode x (* Recursive call: by induction *)
18           ⟨L(true, x, R) * R⟩ (* Commit *)
19         {True}
20     {True}
21 ⟨L(true, x, R) * R⟩
22
23 ⟨R. L(true, x, R) * R⟩
24   let unlockNode x =
25     {True}
26       ⟨R. L(true, x, R) * R⟩ (* Open precondition *)
27       {lk(x) ↦ true * True * R} (* By definition *)
28       lk(x) ← false
29       {lk(x) ↦ false * R}
30       ⟨L(false, x, R)⟩ (* Commit *)
31     {True}
32 ⟨L(false, x, R)⟩
```

Figure 4.7: Proof of the abstract lock specification.

could be locked or unlocked, and returns a locked lock region and the resource R (that can be transferred to the thread's local state). Similarly, the specification of `unlockNode` requires both a locked lock region as well as the resource R, and "puts back" R and unlocks the lock region.

A proof sketch for this specification is given in Figure 4.7. The proof follows the same argument as the proof of the low-level specification in §3.7, except that now we thread the proposition R through the proof. When locking the node, we obtain R from the L predicate, and we return it in the postcondition. When unlocking, we are given R in the precondition, and when we set the lock bit to *false* we put R back into L.

Given this abstract lock region predicate, we can simplify the proof of the single-node template by defining the CSS as:

$$\mathsf{CSS}(r, C) := \exists b. \lceil \tfrac{1}{2} C \rceil^\gamma \, * \, \mathsf{L}\left(b, r, \left(\mathsf{Node}(r, C) \, * \, \lceil \tfrac{1}{2} C \rceil^\gamma \right)\right)$$

By definition of L, this is equivalent to our previous definition of CSS, hence the proof of `cssOp` from §4.3 can be adapted in a straightforward manner to use the high-level lock specification from this section. Going forward, we will use this high-level specification in all template proofs.

CHAPTER 5

The Keyset Resource Algebra

In the previous chapter, we have seen how to use ghost state in order to verify a template algorithm for a single node structure. Extending this proof to structures consisting of multiple nodes needs a notion of the *keyset* of a node, the set of keys for which a node is responsible. In this chapter, we define a keyset resource algebra (RA) that can be used for many single-copy search structures, and demonstrate it by verifying a two-node template.

5.1 A TWO-NODE TEMPLATE

We now look at a search structure that contains two nodes, n_1 and n_2, whose template algorithm is listed in Figure 5.1. Since there are two nodes, the first step in this algorithm is to find the node in which to search for, insert, or delete the given key. This is done via a new helper function findNode. Once the appropriate node n is found, the algorithm proceeds similarly to the single-node template: it locks n, calls decisiveOp on it, and then unlocks it.

Implementations of this template choose not only how to store the keys in a node (e.g., as an array of keys or a list of keys) but also how to divide keys between nodes. For instance, one possible implementation would be to send the odd keys to n_1 and the even keys to n_2. We represent this choice in the template proof via an abstract function[1] $\mathsf{ks}(n)$ that maps a node n to a set of keys we call the *keyset*. Intuitively, we expect the implementation to define the keyset of a node n as the set of keys $\mathsf{ks}(n)$ that, if present in the structure, must be in n. In the above example, $\mathsf{ks}(n_1)$ is the set of odd numbers and $\mathsf{ks}(n_2)$ is the set of even numbers. The proof of the template can use this keyset function to specify the behavior it expects from the findNode helper function:

$$\{\mathsf{True}\}\ \mathsf{findNode}\ n_1\ n_2\ k\ \{n.\ \mathsf{InFP}(n_1, n_2, n) * k \in \mathsf{ks}(n)\}$$

Here, $\mathsf{InFP}(n_1, n_2, n) := (n = n_1 \vee n = n_2)$ is a predicate that captures the fact that n is in the *footprint* of the data structure, i.e., that it is one of the nodes in the data structure.[2] We continue to use the simple spin-lock implementation of lockNode and unlockNode, and use the abstract specifications from Figure 4.6. Note that a thread can call lockNode or unlockNode only on a node x for which it owns the heap cell $\mathsf{lk}(x)$ – this is where the $\mathsf{InFP}(n_1, n_2, n)$ predicate will be used.

---

[1]Abstract functions are like abstract predicates in that the template proof is done without knowing their definition; instead, the proof relies on certain assumptions about them.

[2]While this is a trivial definition for the two-node template, we will use the same predicate to simplify the more complex proofs in later chapters.

```
1 let create () =
2    let n₁, n₂ = allocNodes () in
3    (n₁, n₂)
4
5 let cssOp ω n₁ n₂ k =
6    let n = findNode n₁ n₂ k in
7    lockNode n;
8    let res = decisiveOp ω n k in
9    unlockNode n;
10    res
```

Figure 5.1: A template algorithm for a two-node search structure.

The challenge is in providing a suitable specification for decisiveOp. At the point when decisiveOp is called, only one of the two nodes in the structure is locked by the current thread, and hence any specification for decisiveOp can speak only about the node n. A natural first attempt would be:

$$\{\mathsf{Node}(n, C_n)\} \; \mathtt{decisiveOp} \; \omega \; n \; k \; \{res.\, \mathsf{Node}(n, C'_n) * \Psi_\omega(k, C_n, C'_n, res)\},$$

This spec says decisiveOp converts $\mathsf{Node}(n, C_n)$ (node n with contents C_n) into $\mathsf{Node}(n, C'_n)$ (the node n with updated contents C'_n) such that the search structure specification predicate $\Psi_\omega(k, C_n, C'_n, res)$ holds. However, the trouble is that the postcondition of cssOp requires us to show that the contents of the entire search structure are modified from some C to C' such that $\Psi_\omega(k, C, C', res)$ holds. To complete the proof, we need to show that $\Psi_\omega(k, C_n, C'_n, res) \Rightarrow \Psi_\omega(k, C, C', res)$.

This is not true of arbitrary sets $C_n \subseteq C$ and $C'_n \subseteq C'$. Consider the case where node n_1 has contents $\{1, 3, 8\}$, and n_2 has contents $\{2, 4, 8\}$ and decisiveOp removes key 8 from n_1. Here $C_n = \{1, 3, 8\}$ and $C'_n = \{1, 3\}$, but $C = C' = \{1, 2, 3, 4, 8\}$.

So, we need more constraints. Our example implementation assigned each node a distinct set of keys (n_1 got the odd keys and n_2 got even keys). The missing piece of the proof is the property that the keysets of any two nodes are *disjoint*. If we have a data structure where all keysets are disjoint and the contents of each node n are a subset of the keyset of n, then we can show that it is sufficient for decisiveOp to ensure that Ψ_ω holds on some node n such that $k \in \mathsf{ks}(n)$. We next show how to encode this argument in separation logic using an appropriate resource algebra.

5.2   DISJOINT KEYSETS AND THE KEYSET RA

We define an RA that we use to keep track of the keyset and contents of each node simultaneously:

Definition 5.1   Given a key space $\mathbb{K}$, the *keyset* RA is defined as:

$$\textsc{Keyset} := (\mathbb{K} \times \mathbb{K}) \mid \frac{1}{4} \qquad \mathcal{V}((K, C)) := (C \subseteq K) \qquad \mathcal{V}(\frac{1}{4}) := \mathsf{False}$$

$$(K_1, C_1) \cdot (K_2, C_2) := \begin{cases} (K_1 \cup K_2, C_1 \cup C_2) & \text{if } C_1 \subseteq K_1 \wedge C_2 \subseteq K_2 \wedge K_1 \cap K_2 = \emptyset \\ \frac{1}{4} & \text{otherwise} \end{cases}$$

$$\frac{1}{4} \cdot \_ := \_ \cdot \frac{1}{4} := \frac{1}{4}$$

The unit of this RA is the element $(\emptyset, \emptyset)$.

This is an RA where elements are pairs of sets of keys, where the first set represents the keyset and the second represents the contents of a node (or, more generally, a set of nodes). We also have a special element $\frac{1}{4}$ representing invalid compositions. The validity predicate checks if the contents are a subset of the keyset, and composition is only defined between valid elements whose keysets are disjoint.

In order to use the keyset RA in our proofs, we will need a standard RA construction useful to reason about shared ownership of a logical value. The *authoritative* RA $\textsc{Auth}(M)$ [Iris Team, 2020, Jung et al., 2018], constructed from an arbitrary RA M, is used to model situations where there exists an authoritative element a of M, and threads own fragments b of a such that $b \preccurlyeq a := \exists c. \, a = b \cdot c$.

Definition 5.2   Given an RA M with unit ε, the *authoritative* RA $\textsc{Auth}(M)$ is defined as:

$$\textsc{Auth}(M) := (\mathsf{ex}(M) \mid \frac{1}{4}) \times M \qquad \mathcal{V}((x, b)) := (\exists a. \, x = \mathsf{ex}(a) \wedge b \preccurlyeq a \wedge \mathcal{V}(a))$$

$$(x_1, b_1) \cdot (x_2, b_2) := \begin{cases} (x_1, b_1) & \text{if } x_2 = \mathsf{ex}(\varepsilon) \\ (x_2, b_1) & \text{if } x_1 = \mathsf{ex}(\varepsilon) \\ (\frac{1}{4}, b_1 \cdot b_2) & \text{otherwise} \end{cases}$$

The unit of $\textsc{Auth}(M)$ is the element $(\mathsf{ex}(\varepsilon), \varepsilon)$.

Let $a, b \in M$. When using the $\textsc{Auth}(M)$ RA, we write $\bullet \, a$ for ownership of an authoritative element $(\mathsf{ex}(a), \varepsilon)$ and $\circ \, b$ for fragmental ownership $(\mathsf{ex}(\varepsilon), b)$ and $\bullet \, a, \circ \, b$ for combined ownership $(\mathsf{ex}(a), b)$. The composition operator is defined so that only one authoritative element can be owned, as $(\bullet \, a_1, \circ \, b_1) \cdot (\bullet \, a_2, \circ \, b_2) = (\frac{1}{4}, b_1 \cdot b_2)$ which is invalid. However, multiple fragmental elements can be owned simultaneously, and they compose according to the composition of the underlying RA:

AUTH-FRAG-OP

$$(\circ\, a) \cdot (\circ\, b) = \circ\, (a \cdot b)$$

An important property of authoritative RAs is that if one owns both an authoritative element $\bullet\, a$ and a fragment $\circ\, b$, then by the definition of validity, we know that the fragment is a part of the authoritative element, i.e., $b \preccurlyeq a$.

In our proofs, we will be using AUTH(KEYSET), the authoritative keyset RA. Recall that for the single-node template proof (§4.3), we needed to use the fractional RA so that the shared state could keep track of the global contents even when the single node was locked. Similarly, we need two copies of the ghost state that keeps track of the contents, one copy that always stays in the shared state and is fixed to equal the parameter C in $\mathsf{CSS}(n_1, n_2, C)$, and one copy that is split among nodes and handed out to threads who lock a node. Performing this reasoning with many fractional locations gets messy and tedious, so we instead use the authoritative RA, which was built for such situations.

Using AUTH(KEYSET), we add the formula $\boxed{\bullet(\mathbb{K}, C)}^{\gamma}$ to the definition of CSS to represent the abstract state of the search structure as one whose keyset is the entire key space $\mathbb{K}$ and contains the keys C. Similarly, we represent the local abstract state of a node n by the formula $\boxed{\circ(K_n, C_n)}^{\gamma}$, where K_n and C_n are the keyset and contents, respectively, of n. By the definition of the authoritative RA, the assertion

$$\boxed{\bullet(\mathbb{K}, C)}^{\gamma} * \underset{n \in N}{\text{\Huge $*$}} \boxed{\circ(K_n, C_n)}^{\gamma}$$

expresses that the sets K_n for each $n \in N$ are disjoint and their union is included in $\mathbb{K}$.[3] Moreover, $C_n \subseteq K_n$ and similarly the C_n sets are disjoint and are included in C. If we can associate each C_n and K_n to the contents and keyset, respectively, of n, then an assertion like the one above gives us the desired disjoint decomposition of the abstract state into local states.

---

[3]We cannot use the keyset RA to encode the invariant that the union of the sets K_n *cover* the key space $\mathbb{K}$ because the authoritative RA's validity predicate tells us only that fragmental elements are included in the authoritative element. While practical concurrent search structure implementations will ensure that keysets cover the key space, we do not need to maintain such an invariant in our proofs because we prove only partial correctness, and not termination. For instance, suppose we had a two-node structure where the keysets did not cover the key space. Given a k not in the keyset of either node, the only way for findNode n_1 n_2 k to satisfy its specification is for it to not terminate. See §14.2.1 for a discussion on how to extend our proofs to verify termination.

The AUTH(KEYSET) RA has frame-preserving updates such as the following, which we will use to update the ghost state when we insert or delete a key k:

KS-INS

$$\frac{k \notin K_n}{\bullet(K, C), \circ(K_n, C_n) \rightsquigarrow \bullet(K, C \cup \{k\}), \circ(K_n, C_n \cup \{k\})}$$

KS-DEL

$$\frac{k \in K_n}{\bullet(K, C), \circ(K_n, C_n) \rightsquigarrow \bullet(K, C \setminus \{k\}), \circ(K_n, C_n \setminus \{k\})}$$

For example, KS-DEL says that if $\overline{\bullet(K, C)}^\gamma$ and $\overline{\circ(K_n, C_n)}^\gamma$ are valid resources such that $k \in K_n$ then we can update the fragment to $(K_n, C_n \setminus \{k\})$ (for instance when we remove k from the contents of a node n) and the authoritative resource to $(K, C \setminus \{k\})$ (meaning k is also removed from the global contents). Combining this with KS-INS for insertions, we get the following lemma:

KS-UPD

$$\overline{\bullet(K, C)}^\gamma * \overline{\circ(K_n, C_n)}^\gamma * k \in K_n * \Psi_\omega(k, C_n, C_n', res)$$
$$\Rightarrow \exists C'. \overline{\bullet(K, C')}^\gamma * \overline{\circ(K_n, C_n')}^\gamma * \Psi_\omega(k, C, C', res)$$

The lemma KS-UPD captures the intuition that changes made locally to C_n percolate through the global contents C due to disjointness of keysets and the fact that a node's contents is always a subset of its keyset. The lemma is proved from KS-INS and KS-DEL by case analysis on the operation ω and application of rules GHOST-UPDATE, VS-TRANS, and VS-FRAME.

5.3 PROOF OF THE TWO-NODE TEMPLATE

We can now prove the two-node template (Figure 5.2). The definition of CSS has been extended to account for two nodes and a lock region for each. It also contains the authoritative version of the keyset and global contents: $\overline{\bullet(\mathbb{K}, C)}^\gamma$. Each node is represented by the node predicate $N(n)$, which contains the abstract predicate $\mathsf{Node}(n, C_n)$ that is implementation-specific as well as the fragment containing n's keyset and contents $\overline{\circ(\mathsf{ks}(n), C_n)}^\gamma$.

We first describe the proof of the `create` method that constructs the search structure. The specification of `create` is a Hoare triple because the search structure is created before the concurrent context begins or any threads are created. We use the helper function `allocNodes` that allocates the nodes $\mathsf{Node}(n_1, \emptyset)$ and $\mathsf{Node}(n_2, \emptyset)$ with empty contents. It also creates the lock bit for each node. We then use GHOST-ALLOC to allocate a ghost location γ with contents $\bullet(\mathbb{K}, \emptyset) \cdot \circ(\mathsf{ks}(n_1), \emptyset) \cdot \circ(\mathsf{ks}(n_2), \emptyset)$. This ghost state is valid because the keysets of the two nodes are disjoint and included in the key space $\mathbb{K}$. We then use GHOST-OP to split this into the authoritative version and two fragments, which we then fold into the predicates $N(n_1)$ and $N(n_2)$.

1 $\mathsf{InFP}(n_1, n_2, n) := (n = n_1 \vee n = n_2)$

2 $\mathsf{N}(n) := \exists C_n.\ \mathsf{Node}(n, C_n) * \overline{\lfloor \circ(\mathsf{ks}(n), C_n) \rfloor}^{\gamma}$

3 $\mathsf{CSS}(n_1, n_2, C) := \exists b_1, b_2.\ \overline{\lfloor \bullet(\mathbb{K}, C) \rfloor}^{\gamma} * \mathsf{L}(b_1, n_1, \mathsf{N}(n_1)) * \mathsf{L}(b_2, n_2, \mathsf{N}(n_2))$

4

5 $\{\mathsf{True}\}$
6 **let** create () =
7    $\{\mathsf{True}\}$
8   **let** $n_1,\ n_2$ = allocNodes () **in**
9    $\{\mathsf{Node}(n_1, \emptyset) * \mathsf{lk}(n_1) \mapsto \mathit{false} * \mathsf{Node}(n_2, \emptyset) * \mathsf{lk}(n_2) \mapsto \mathit{false}\}$
10   $\{\mathsf{Node}(n_1, \emptyset) * \mathsf{lk}(n_1) \mapsto \mathit{false} * \mathsf{Node}(n_2, \emptyset) * \mathsf{lk}(n_2) \mapsto \mathit{false} * \overline{\lfloor \bullet(\mathbb{K}, \emptyset) \cdot \circ(\mathsf{ks}(n_1), \emptyset) \cdot \circ(\mathsf{ks}(n_2), \emptyset) \rfloor}^{\gamma}\}$
11   $\{\overline{\lfloor \bullet(\mathbb{K}, \emptyset) \rfloor}^{\gamma} * \mathsf{L}(\mathit{false}, n_1, \mathsf{N}(n_1)) * \mathsf{L}(\mathit{false}, n_2, \mathsf{N}(n_2))\}$
12   $\{\mathsf{CSS}(n_1, n_2, \emptyset)\}$
13   $(n_1,\ n_2)$
14 $\{\mathsf{CSS}(n_1, n_2, \emptyset)\}$

15

16 $\langle C.\ \mathsf{CSS}(n_1, n_2, C) \rangle$
17 **let** cssOp ω n_1 n_2 k =
18   $\{\mathsf{True}\}$
19   **let** n = findNode n_1 n_2 k **in**
20   $\{\mathsf{InFP}(n_1, n_2, n) * k \in \mathsf{ks}(n)\}$
21     $\langle \mathsf{CSS}(n_1, n_2, C) * \mathsf{InFP}(n_1, n_2, n) * k \in \mathsf{ks}(n) \rangle$
22   lockNode n;
23     $\langle \mathsf{CSS}(n_1, n_2, C) * \mathsf{N}(n) * k \in \mathsf{ks}(n) \rangle$
24   $\{\mathsf{N}(n) * k \in \mathsf{ks}(n)\}$
25   $\{\mathsf{Node}(n, C_n) * \overline{\lfloor \circ(\mathsf{ks}(n), C_n) \rfloor}^{\gamma} * k \in \mathsf{ks}(n)\}$
26   **let** res = decisiveOp ω n k **in**
27   $\{\mathsf{Node}(n, C_n') * \overline{\lfloor \circ(\mathsf{ks}(n), C_n) \rfloor}^{\gamma} * \Psi_\omega(k, C_n, C_n', \mathit{res}) * k \in \mathsf{ks}(n)\}$
28     $\langle \mathsf{Node}(n, C_n') * \overline{\lfloor \circ(\mathsf{ks}(n), C_n) \rfloor}^{\gamma} * \Psi_\omega(k, C_n, C_n', \mathit{res}) * \overline{\lfloor \bullet(\mathbb{K}, C) \rfloor}^{\gamma} * k \in \mathsf{ks}(n) * \cdots \rangle$
29     $\langle \mathsf{Node}(n, C_n') * \overline{\lfloor \circ(\mathsf{ks}(n), C_n') \rfloor}^{\gamma} * \Psi_\omega(k, C, C', \mathit{res}) * \overline{\lfloor \bullet(\mathbb{K}, C') \rfloor}^{\gamma} * \cdots \rangle$ (* By KS-UPD *)
30   unlockNode n;
31     $\langle \mathsf{CSS}(n_1, n_2, C') * \Psi_\omega(k, C, C', \mathit{res}) \rangle$
32   $\{\mathsf{True}\}$
33   res
34 $\langle v.\ \mathsf{CSS}(n_1, n_2, C') * \Psi_\omega(k, C, C', v) \rangle$

Figure 5.2: Proof of the two-node template algorithm.

$$\{\mathsf{True}\} \ \mathtt{allocNodes} \ () \ \{n_1, n_2. \ \mathsf{Node}(n_1, \emptyset) * \mathsf{lk}(n_1) \mapsto \mathit{false} * \mathsf{Node}(n_2, \emptyset) * \mathsf{lk}(n_2) \mapsto \mathit{false}\}$$

$$\{\mathsf{True}\} \ \mathtt{findNode} \ n_1 \ n_2 \ k \ \{n. \ \mathsf{InFP}(n_1, n_2, n) * k \in \mathsf{ks}(n)\}$$

$$\{\mathsf{Node}(n, C_n)\} \ \mathtt{decisiveOp} \ \omega \ n \ k \ \{\mathit{res}. \ \mathsf{Node}(n, C'_n) * \Psi_\omega(k, C_n, C'_n, \mathit{res})\}$$

$$\mathsf{Node}(n, C_n) * \mathsf{Node}(n, C'_n) \ \ast\!\!\!-\!\!\ast \ \mathsf{False}$$

Figure 5.3: The assumptions made by the two-node template on implementations.

We can then combine these with the lock locations to get the proof context shown in line 11. By definition of CSS, this gives us the desired postcondition, which is a search structure with empty contents.

Moving to cssOp, the call to findNode is handled as explained previously, using the specification given in Figure 5.3. To prove the precondition of lockNode, we open the precondition and use the predicate $\mathsf{InFP}(n_1, n_2, n)$ that we obtained from findNode to show that we own $\mathsf{L}(b, n, \mathsf{N}(n))$. After lockNode, we can move the predicate $\mathsf{N}(n)$ from the shared state into our local state as before. We then use the specification of decisiveOp to get a modified node predicate $\mathsf{Node}(n, C'_n)$ and $\Psi_\omega(k, C_n, C'_n, \mathit{res})$.

As with the single-node template, the linearization point is at the call to unlockNode. We use the rule AU-COMMIT to open the precondition and get access to the shared state, obtaining the resources shown in the intermediate assertion on line 28. We then use KS-UPD to update both the node's fragment of the keyset RA as well as the authoritative element to the new contents, and obtain the resource $\Psi_\omega(k, C, C', \mathit{res})$. This step corresponds to the reasoning that since the decisive operation was performed on a node n such that $k \in \mathsf{ks}(n)$, the global contents also change appropriately. We can then apply unlockNode's specification to change the lock location of n, and return $\mathsf{N}(n)$ to the shared state, obtaining the postcondition $\mathsf{CSS}(n_1, n_2, C') * \Psi_\omega(k, C, C', \mathit{res})$.

CHAPTER 6

The Edgeset Framework for Single-Copy Structures

This chapter introduces the *edgeset framework* that allows one to view a single-copy search structure as an abstract graph whose nodes are labeled by their contents and edges are labeled by *edgesets*. Such an abstract view allows us to define template algorithms for concurrent search structures that fix a concurrent technique but can be instantiated to multiple concrete data structures. Edgesets, and template algorithms based on them, were first introduced by Shasha and Goodman [1988].

We first present the intuition behind the fundamental concept of an edgeset using a library analogy and show that it applies across existing search structures. Next, we describe the B-link tree data structure, a highly efficient and popular algorithm that uses the *link* technique of synchronization. We then apply the edgeset notion to derive a template algorithm that can be instantiated to any concurrent search structure algorithm that supports a link-based redirection for operations that arrive at an incorrect node, including the B-link tree. Finally, we show how edgesets can be used to define the keyset of each node, thereby allowing us to extend the proof strategy of the two-node template from Chapter 5 to templates with arbitrarily many nodes.

6.1   AN INTUITIVE INTRODUCTION TO EDGESETS

Every search structure supports the notion of navigation. This implies that searches (and the search portion of inserts, deletes, and updates) follow edges (normally in the form of either explicit or implicit pointers) as they traverse a path to an appropriate destination node. For example, if a search on a binary search tree for key 5 arrives at a node n having value 7, then the search will proceed to the left child of n. Similarly, a search for key 10 on a hash structure characterized by hash function h will proceed to the node $h(10)$.

In our framework, we associate each edge (n, n') of a search structure with a set of key values called the *edgeset*, written $\text{es}(n, n')$. If a key k belongs to $\text{es}(n, n')$, then a search that arrives at n will proceed to n'. In a sorted linked list (Figure 6.1, left), if node n has a key 6, then $\text{es}(n, n.\text{next})$ consists of all values greater than 6. In a binary search tree (Figure 6.1, middle), if node n has the key value 2, then $\text{es}(n, n.\text{left})$ consists of all values less than 2. Similarly, for a hash structure having hash function h and a bucket i, the edgeset from the root of the hash structure to i consists of $\{k \mid h(k) = i\}$ (Figure 6.1, right).

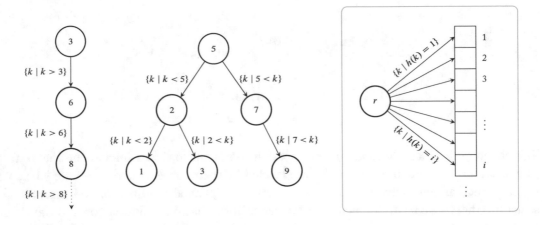

Figure 6.1: Examples of edgesets (shown as edge labels) in a sorted linked list (left), a binary search tree (middle), and a hash structure (right).

Returning to the library of books from the introduction, the card catalog may provide an edgeset to a shelf n consisting of all the books whose first authors' last names begin with P. When Alice moves some of the books from n to n', then her sticky note should correspond to the edgeset of the books that have moved, e.g., books whose first authors' last names lie in the range Po ... Pz. When Alice later changes the card catalog to create entries for the books that have moved to n', she is increasing $es(cat, n')$ from the empty set to the range Po ... Pz and decreasing $es(cat, n)$ from P to Pa ... Pn.

Thus, the edgeset concept gives us the means to express a search/insert/update/delete algorithm that applies to any search structure. A search for key k starts at a root (it is fine for search structures to have many roots) and stops at node n if no edgeset leaving n contains k.

On well-formed search structures, edgesets have the following two properties:

1. For every node n in a search structure, the edgesets leaving n are mutually exclusive. This ensures that searches have a unique next node to navigate to.

2. The set of keys in n, denoted $cnts(n)$ (for contents), will be disjoint with the edgesets leaving n. This ensures that a search for k will not leave node n when k is in node n.

In a binary search tree, for example, the edgesets of the left and right children of node n are disjoint and are of course disjoint from the key in node n.

6.2   B-LINK TREES

The B-link tree (Figure 6.2) is an implementation of a concurrent search structure based on the B-tree. A B-tree is a generalization of a binary search tree, in that a node can have more than two children. In a binary search tree, each node contains a key k_0 and up to two pointers y_l and y_r. An operation on k takes the left branch if $k < k_0$ and the right branch otherwise. A B-tree generalizes this by having l sorted keys $k_0, \ldots, k_{l-1}$ and $l + 1$ pointers $y_0, \ldots, y_l$ at each node, such that $B \leq l + 1 < 2B$ for some constant B. At internal nodes, an operation on k takes the branch y_i if $k_{i-1} \leq k < k_i$.

In the most common implementations of B-trees (called B+ trees), the keys are stored only in leaf nodes; internal nodes contain "separator" keys for the purpose of routing only, and therefore are not part of the contents of the structure. For example, the search structure depicted in Figure 6.2 (bottom) has key contents $\{1, 2, 4, 6, 7, 8, 9\}$. When an operation arrives at a leaf node n, it proceeds to insert, delete, or search for its operation key in the keys of n. To avoid interference, each node has a lock that must be held by an operation before it reads from or writes to the node.

When a node n becomes full, a separate maintenance thread performs a split operation by transferring half its keys (and pointers, if it is an internal node) into a new node n', and adding a link to n' from the parent of n. A concurrent algorithm needs to ensure that this operation does not cause concurrent operations at n looking for a key k that was transferred to n' to conclude that k is not in the structure. The B-link tree solves this problem by linking n to n' and storing a key k' (the key in the gray box in the figure) that indicates to concurrent operations that the key k can be reached by following the link edge if $k > k'$. (That is what we mean by the intuitive notion of redirection above.)

To reduce the time the parent node is locked, this split is performed in two steps: (i) a half-split step that locks n, transfers half the keys to n', and adds a link from n to n' and (ii) a complete-split performed by a separate thread that takes a half-split node n, locks the parent of n, and adds a pointer to n'.

Figure 6.2 (top) shows the state of a B-link tree where node n has become full. We thus perform a half-split that moves its children $\{y_2, y_3\}$ to a new node n' and adds a link edge from n to n'. The key 5 in the gray box in n directs operations on keys $k \geq 5$ via the link edge to n'. The bottom figure shows the state after this half-split but before the complete-split when the pointer of n' will be added to r (shown using a dotted edge in Figure 6.2).

6.3   ABSTRACTING SEARCH STRUCTURES USING EDGESETS

The link technique is not restricted to B-trees: consider a hash table implemented as an array of pointers, where the ith entry refers to a bucket node that contains an array of keys $k_0, \ldots, k_l$ that all hash to i. When a node n gets full, it is locked, its keys are moved to a new node n' with

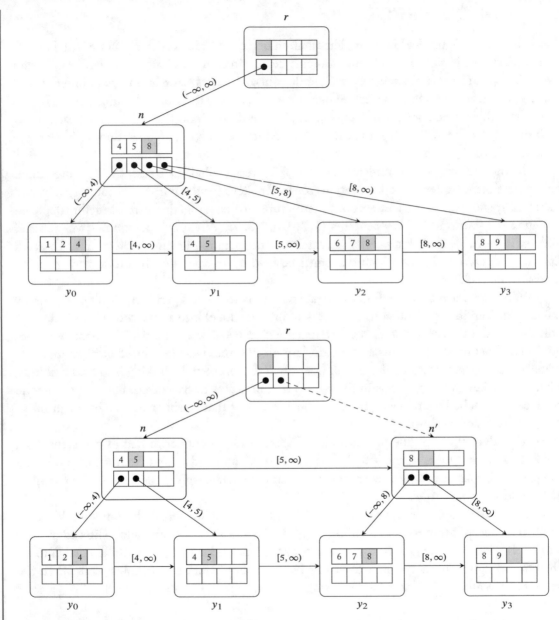

Figure 6.2: A B-link tree before (top) and after (bottom) a half-split on node n, which was full, that transferred children y_2 and y_3 to a new node n'. A subsequent complete-split will add n' to the parent r (the dashed edge). Each node contains an array of keys (top array) and an array of pointers (bottom array), and a separator value in a gray box directing operations on larger keys to follow the link edge to the right-neighboring node. Edges are labeled by their edgeset (§6.3).

twice the space, and n is linked to n'. Again, a separate operation locks the main array entry and updates it from n to n'. Thus, B-link trees and hash-link structures follow the same principle as Alice in the book library.

While these data structures look completely different, the main operations of search, insert, and delete follow the same abstract algorithm. In both B-link trees and hash-link structures, there is some local rule by which operations are routed from one node to the next, and both introduce link edges when keys are moved to ensure that no other operation loses its way.

Reprising the discussion of §6.1, we view the state of a search structure abstractly as a mathematical graph. Each node in this graph can represent anything from two adjacent heap cells (in the case of a singly-linked list) to a collection of arrays and fields (in the case of a B-tree), and this mapping is determined by the specific implementation under consideration. We then define the *edgeset* of an edge (n, n'), written $\mathsf{es}(n, n')$, to be the set of operation keys for which an operation arriving at a node n traverses (n, n').

The B-link tree in Figure 6.2 labels each edge with its edgeset; the edgeset of (n, y_1) is $[4, 5)$ and the edgeset of the link edge (y_0, y_1) is $[4, \infty)$. Note that 4 is in the edgeset of (y_0, y_1) even though an operation on 4 would not normally reach y_0. This is deliberate. In order to make edgeset a local quantity, we say $k \in \mathsf{es}(n, n')$ if an operation on k would traverse (n, n') assuming it somehow found itself at n.

In the hash table, assuming there exists a global root node, the edgeset from the root to the ith array entry is $\{k \mid h(k) = i\}$, i.e., all the key values for which a search would go to the node of the ith array entry. By contrast, the edgeset from an array entry to the bucket node is the set of all keys $\mathbb{K}$, as is the edgeset from a deleted bucket node to its replacement. (The reason is that once we arrive at an array entry (or a deleted node), we follow the outgoing edge no matter which key we are looking for.)

6.4   THE LINK TEMPLATE

Figure 6.3 lists the link template algorithm [Shasha and Goodman, 1988] that uses edgesets to describe the algorithm used by all core operations for both B-link trees and hash tables in a uniform manner. The algorithm is described in the ML-like language that we use throughout the monograph, and is described in more detail in §2.2. The algorithm assumes that an implementation provides certain primitives or helper functions that satisfy certain properties. For example, findNext is a helper function that finds the next node to visit given a current node n and an operation key k, by looking for an edge (n, n') with $k \in \mathsf{es}(n, n')$. For the B-link tree, findNext does a binary search on the keys in a node to find the appropriate pointer to follow. For the hash table, when at the root, findNext returns the edge to the array element indexed by the hash of the key; when at a bucket node n, findNext returns the link edge if it exists and k is not in n. The function cssOp can be used to build implementations of all three search structure operations by implementing the helper function decisiveOp to perform the desired operation (read, add, or remove) of key k on the node n.

```
1 let create () =                      12 let rec cssOp ω r k =
2   let r = allocRoot () in            13   let n = traverse r k in
3   r                                  14   match decisiveOp ω n k with
4                                      15   | None -> unlockNode n;
5 let rec traverse n k =               16       cssOp ω r k
6   lockNode n;                        17   | Some res -> unlockNode n;
7   match findNext n k with            18       res
8   | None -> n
9   | Some n' ->
10      unlockNode n;
11      traverse n' k
```

Figure 6.3: The link template algorithm. The cssOp method is the main method, and represents the core search structure operations (search, insert, and delete) via the parameter ω. It uses an auxiliary method traverse that recursively traverses the search structure until it finds the node upon which to operate (the node containing k in its keyset, as described in §6.5). This template can be instantiated to the B-link tree algorithm by providing implementations of helper functions findNext and decisiveOp. findNext $n\,k$ returns Some n' if $k \in$ es(n, n') and None if there exists no such n'. decisiveOp $n\,k$ performs the operation ω (either search, insert, or delete) on k at node n.

An operation on key k starts at the root r, and calls a function traverse on line 13 to find the node on which it should operate. traverse is a recursive function that works by following edges whose edgesets contain k (using the helper function findNext on line 7) until the operation reaches a node n with no outgoing edge having an edgeset containing k (so $k \in$ ks(n)). Note that the operation locks a node only during the call to findNext, and holds no locks when moving between nodes. traverse terminates when findNext does not find any n' such that $k \in$ es(n, n'). In the B-link tree example, this corresponds to finding the appropriate leaf.

At this point, the thread performs the decisive operation on n (line 14). Note that for the link template, in contrast to the single-node template from Figure 2.1, we assume decisiveOp returns an optional Boolean value so that it can signal when it fails. If it is not possible for the decisive operation to be completed because, say, an insert operation encounters a full node, decisiveOp returns None and the algorithm unlocks n, gives up, and starts from the root again. Note that non-completion does not imply the existence of race conditions or null pointer exceptions. Our proofs show that such errors are impossible. If the decisive operation can be completed (i.e., it succeeds), then decisiveOp returns Some res and the algorithm unlocks n and returns res.

If we can verify this link template algorithm with a proof that is parameterized by the helper functions, then we can reuse the proof across diverse search structures.

6.5   FROM EDGESETS TO KEYSETS

In this section, we describe the high-level argument behind our single-copy proofs. The proof of the two-node template in Chapter 5 used the keyset RA with a fixed keyset function (e.g., odd keys in one node and even keys in another). For more realistic data structures with a dynamic and unbounded number of nodes, the keyset function cannot be determined statically ahead-of-time. We use the edgeset framework to define the keyset of each node using the edgesets and the graph structure of the underlying state.

We once again work on the abstract graph view of the search structure from §6.1. Let the union of edgesets leaving a node n be the *outset* $\mathsf{outs}(n)$. Suppose we denote the intersection of edgesets along the path from the root (there can be many roots, but for now let us consider just one) leading to a given node n as the *inset* of n, denoted $\mathsf{ins}(n)$. We can then define the *keyset* of n as:

$$\mathsf{ks}(n) := \mathsf{ins}(n) \setminus \mathsf{outs}(n)$$

For example, if a node n in a binary search tree has key 7 (see Figure 6.1), the parent of n (which is also the root) has key 5, and n has a right child but no left child, then $\mathsf{ks}(n)$ are keys greater than 5 and less than or equal to 7, $\{k \mid 5 < k \leq 7\}$. Of course, the contents of n, namely 7, is in that set.

Sequential search structure algorithms will ensure the following two invariants:

1. For every node n, the contents $\mathsf{cnts}(n)$ are a subset of the keyset $\mathsf{ks}(n)$.

2. The keysets of every pair of distinct nodes are disjoint.

We call these the *keyset invariants*.[1] In our proofs, we encode these invariants using the keyset RA from Chapter 5. The algebraic properties of the keyset RA will ensure that these invariants are maintained, and the edgeset framework can be used to relate the ghost state to the concrete contents of the search structure.

All sequential single-copy search structures maintain the keyset invariants. A consequence of these invariants is that a search for key k can examine the node n such that $k \in \mathsf{ks}(n)$ to determine whether k is anywhere in the search structure (because the keyset invariants implies that k cannot be anywhere else). That examination of n is called the *decisive operation* of the search. For many structures of interest, the decisive operations of insert, delete, and update of key k would also apply to the node n having k in its keyset or to a node linked from n.

Maintaining the keyset invariants in the concurrent setting may require more metadata than in the sequential setting. In our library example, if Alice were sure there were no patrons in the library whenever she reshelves books, then she could move books from n to n' and adjust the card catalog without ever leaving a note in n. In the concurrent setting, if she did not leave

---

[1] Note that we do not require the invariant that the union of keysets of all nodes cover the keyspace $\mathbb{K}$; as discussed in §5.2, this is only needed to prove termination.

the note, then at some point $cnts(n')$ would not be a subset of $ks(n')$ with the consequence that a concurrent patron might fail to find a book that is sitting on shelf n'.

For multicopy search structures such as log structured merge (LSM) trees, the edgesets leaving a node are disjoint and are also disjoint from the contents of that node. The algorithms establish a condition called *search recency*. We will explain how this works in the multicopy chapters starting with a slightly different library analogy again in Chapter 9.

CHAPTER 7

The Flow Framework

The preceding chapters have described how to use separation logic and ghost state to verify template algorithms for one and two-node structures. Most useful real-world structures, like the B-link tree (Chapter 6), have a dynamic and unbounded number of nodes. Extending our proofs to structures with unbounded nodes presents a challenge, because we want to prove that a thread preserves global invariants while reasoning only about the few nodes that are accessed or modified by an operation.

The *flow framework* [Krishna et al., 2018, 2020c] is a systematic proof methodology for local reasoning about global graph properties. The flow framework has been applied to perform proofs of a variety of data structures, sequential and concurrent, and the framework's meta theory has been formalized and machine checked in both Coq [Krishna et al., 2020b] and Isabelle [Pöttinger, 2020].

This chapter motivates the flow framework and presents the key aspects of the framework that we use in this monograph. We then present how to encode keysets using flows, thereby allowing us to reason locally about modifications to the contents or keysets of individual nodes.

7.1 MOTIVATION

To see why the two-node template proof from §5.3 does not extend to the unbounded-node setting, recall that we represented the shared state using the CSS predicate, defined as:

$$\mathsf{N}(n) := \exists C_n.\ \mathsf{Node}(n, C_n) * \boxed{\circ(\mathsf{ks}(n), C_n)}^{\gamma}$$
$$\mathsf{CSS}(n_1, n_2, C) := \exists b_1\, b_2.\ \boxed{\bullet(\mathbb{K}, C)}^{\gamma} * \mathsf{L}(b_1, n_1, \mathsf{N}(n_1)) * \mathsf{L}(b_2, n_2, \mathsf{N}(n_2))$$

Extending CSS to describe an unbounded number of nodes can be done using the iterated separating conjunction as (where r is the root node):

$$\mathsf{CSS}(r, C) := \exists N.\ \boxed{\bullet(\mathbb{K}, C)}^{\gamma} * \underset{n \in N}{\scalebox{1.5}{$*$}} (\exists b.\ \mathsf{L}(b, n, \mathsf{N}(n)))$$

This formula says there exists a set of nodes N, where each $n \in N$ has a lock location set to some Boolean b and the resources in $\mathsf{N}(n)$. Extending the definition of N, on the other hand, is trickier. Recall that $\mathsf{ks}(n)$ was an abstract function, whose definition was to be provided by the implementation. For the two-node template, our example implementation used the following definition of the keyset:

$$\mathsf{ks}(n) := \begin{cases} \{k \mid k \text{ is odd}\} & n = n_1 \\ \{k \mid k \text{ is even}\} & \text{otherwise} \end{cases}$$

Unfortunately, the keyset in most search structures is a *global* quantity, i.e., $\mathsf{ks}(n)$ depends on nodes other than just n. For example, the rules of a B-tree dictate that the keyset of a node n depends on the keys in all the nodes on the path from the root to n. In Figure 6.2, the keyset of y_0 is $(-\infty, 4)$, and the keyset of y_2 is $[5, 8)$. This makes the keyset a function that depends on the entire data structure. Consequently, every time an operation modifies a single node or edge, the keyset function could potentially change. The proof of every operation would then have to reason about the change it makes to the keyset of every node in the structure, which is not a scalable approach. We thus need a way to reason about a global graph quantity such as the keyset in a local manner.

7.2   LOCAL REASONING ABOUT GLOBAL PROPERTIES

As noted above, while separation logic is based on the concept of *local reasoning*, many important properties of data structure graphs depend on non-local information. For instance, we cannot express the property that a graph is a tree by conjoining per-node invariants. As we have seen above, we also cannot write down the keyset of a B-tree as a local function of each node. The flow framework is a separation logic based approach that provides a mechanism to reason about global quantities in local proofs.

The flow framework uses the concept of a *flow*—a function from nodes to values from some *flow domain*—to specify global graph properties in terms of node-local invariants. These flow values must satisfy the *flow equation*, i.e., they must be a fixpoint of a set of algebraic equations induced by the entire graph (thereby allowing one to capture global constraints at the node level). When modifying a graph, the framework allows one to perform a local proof that flow-based invariants are maintained via the notion of a *flow interface*. This is an abstraction of a graph region that specifies the flow values entering and exiting the region; if this interface is preserved then the flow values of the rest of the graph will be unchanged.

The rest of this section illustrates these concepts by considering the simple example of a linked-list data structure. We will come back to search structures in §7.4.

Suppose we have a graph G on a set of nodes N and we want to express the global property that it is a list rooted at some node r in terms of a condition on each node. To do this, we need to know some global information at each node: for instance, suppose there existed a function pc that mapped each node n to the number of paths from r to n.[1] If for every node n, $\mathsf{pc}(n) = 1$ and n has at most one outgoing edge (both node-local assertions) then we know that G must be a list rooted at r.

---

[1]We assume a definition of pc where $\mathsf{pc}(r) = 1$ even in acyclic graphs, because typically we are interested in the reachability of nodes in the data structure from an external pointer.

This path-counting function pc is an example of a flow because it can be defined as a solution to the flow equation:

$$\forall n \in N. \, fl(n) = in(n) + \sum_{n' \in N} e(n', n)\big(fl(n')\big) \qquad \text{(FlowEqn)}$$

This is a fixpoint equation on a function $fl\colon N \to M$, where M is a flow domain, in is an *inflow* that specifies the default/initial flow value of each node, and e is a mapping from pairs of nodes to *edge functions* that determine how the flow of one node affects the flow of its neighbor. We define flow domains formally below, but for now we can think of them as monoids with a commutative operator $+$.

The flow framework works with directed partial graphs that are augmented with a flow, called flow graphs. A flow graph is a tuple $H = (N, e, fl)$ consisting of a finite set of nodes $N \subseteq \mathfrak{N}$ ($\mathfrak{N}$ is potentially infinite), a mapping from pairs of nodes to edge functions $e\colon N \times \mathfrak{N} \to E$ (where E is a set of functions $M \to M$), and a function fl such that (FlowEqn) is satisfied for some inflow in. Flow graph composition $H_1 \odot H_2$ is a partial operator that is a disjoint union of the nodes, edges, and flow values and is defined only if the resulting graph continues to satisfy (FlowEqn).

In the case of the path-counting flow, the flow domain M is $\mathbb{N}$ and the inflow is $in(n) := (n = r \, ? \, 1 : 0)$ (the root has inflow of 1 and all other nodes have inflow of 0). The edge function is

$$e(n, n') := \begin{cases} \lambda_{\text{id}} & (n, n') \in G \\ \lambda_0 & \text{otherwise,} \end{cases}$$

where $\lambda_{\text{id}} := (\lambda x. \, x)$ is the identity function and $\lambda_0 := (\lambda x. \, 0)$ is the identically zero function. This means pairs of nodes (n, n') that have an edge between them propagate the path count from n to n', and propagate a path count of 0 if there is no edge between them. The flow equation then reduces to the familiar constraint that the number of paths from r to n, $fl(n)$, equals 1 if $n = r$ else 0, plus the sum of the number of paths to all n' that have an edge to n.

The problem with assuming that each node knows a flow value that satisfies some global constraint over the entire graph is that when a program modifies the graph, it can be hard to show that the flow-based invariants are maintained. In particular, when the program modifies a small part of the graph, say by modifying a single edge, we would like to prove that the flow invariants are preserved by reasoning only about a small region around the modified edge. The flow framework enables such local proofs by means of an abstraction of flow (sub)graphs called flow interfaces.

Consider the simple example of a singly linked list deletion procedure that unlinks[2] a given node n from the list (Figure 7.1). The program swings the pointer from n's predecessor l to n's successor m. We use the path-counting flow and the flow-based local constraints described above to express the invariant that the graph is a list (we show how to formally express this later).

---

[2]Recall from Chapter 3 that we assume a garbage-collected setting in this monograph.

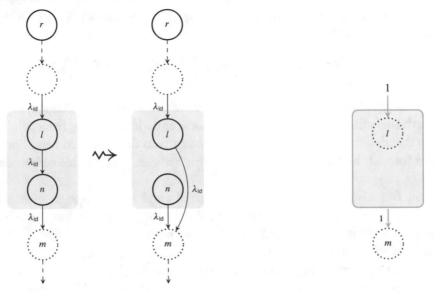

Figure 7.1: A flow-based view of the delete operation on a linked-list. The operation unlinks a node n from a linked-list by swinging the pointer from its predecessor l to its successor m. Edges are labeled with the path-counting flow domain (λ_0 edges omitted). The interface of the blue region $\{l, n\}$ is shown on the right, and is preserved by this update.

For a flow graph H over the path-counting flow domain, modifying a single edge (n, n') can potentially change the flow (the path-count) of every node reachable from n. However, notice that the modification shown in Figure 7.1 changes (l, n) to (l, m) where m is the successor of n. This preserves the flow of every node outside the modified subgraph $H_1 = H|_{\{l,n\}}$ (shown in blue in Figure 7.1) because there was one path coming out of H_1 and one path going to m both before and after the modification.

Flow interfaces build on this intuition; the interface $I = (in, out)$ of a flow (sub)graph H with domain N is a tuple consisting of the inflow $in: N \to M$ and the outflow $out: (\mathfrak{N} \setminus N) \to M$. For the linked list deletion example, the interface of H_1 (the subgraph of nodes $\{l, n\}$ on the left of Figure 7.1) is $I_1 = (in_1, out_1) = (\{l \rightarrowtail 1, n \rightarrowtail 0\}, \lambda_0[m \rightarrowtail 1])$. This tells us that H_1 has one incoming path to l (and none to n) from the rest of the graph, and that there is one one outgoing path from H_1 to m. The interface of H_1', the subgraph after modification shown in the center of Figure 7.1, is also $(\{l \rightarrowtail 1, n \rightarrowtail 0\}, \lambda_0[m \rightarrowtail 1])$. This common interface is depicted abstractly on the right of the figure.

The notion of inflow here generalizes that from (FlowEqn) (the $in(n)$ term) to a setting where we restrict our attention to an arbitrary subgraph H'. Intuitively, for each node $n \in H'$ the inflow $in(n)$ is the number of incoming paths from nodes outside H' directly to n. When considering the "entire" graph, as in (FlowEqn), we think of the inflow *in* to be a map that identifies the root or roots from which we start counting paths (this is why we used $in(n) := (n = r \,?\, 1 : 0)$ above).

Note that the inflow is not the same as the flow. The inflow of n with respect to H_1 is 0 because there are no paths coming from outside of H_1 directly to n. By contrast, the flow of n is 1, because there is one path to n coming through l in H_1. The flow is the number of paths to n, but the inflow depends on the subgraph in question and is the number of paths to n from outside this subgraph coming directly to n. Similarly, for each external node $n' \notin H'$, $out(n')$ is the number of outgoing paths from H'.

Formally, the inflow of $H = (N, e, fl)$ is the *in* that satisfies (FlowEqn) (this is unique [Krishna et al., 2020c]) and the outflow is defined as $out(n) := \sum_{n' \in N} e(n', n)(fl(n'))$.

The flow framework tells us that if we have $H = H_1 \odot H_2$ and we modify H_1 to some H_1' with the same interface, then $H' = H_1' \odot H_2$ exists. This means that the flow of all nodes in H_2 is unchanged; thus it suffices to check that H_1' satisfies the flow-based invariant and has the same interface as H_1, which are both local checks.

The invariants of a wide variety of algorithms have been described using flows, including overlaid data structures (trees overlaid with lists such as B-link trees), graph algorithms (Dijkstra's shortest path algorithm, the priority inheritance protocol), and object-oriented design patterns (subject/observer, composite design pattern). In fact, one can capture any graph property of interest by instantiating the flow domain appropriately [Krishna et al., 2020c]. This involves specifying the desired property in terms of two constraints: a local constraint on the flow of every node, and a global constraint on the inflow and outflow of the entire data structure (the latter is typically used to identify root nodes). We will demonstrate this in §7.4 when we encode keysets using flows. First, we show how to perform flow-based reasoning in Iris.

7.3   THE FLOW INTERFACE RESOURCE ALGEBRA

This section formally defines the notions presented in the previous section, in particular, a flow interface resource algebra that will allow us to perform flow-based reasoning in Iris.

Definition 7.1   Flow Domain. A *flow domain* $(M, +, 0, E)$ consists of a commutative cancellative (total) monoid $(M, 0, +)$ and a set of functions $E \subseteq M \to M$.

Example 7.2   The flow domain used for the path-counting flow is $(\mathbb{N}, +, 0, \{\lambda_{id}, \lambda_0\})$, consisting of the monoid on natural numbers under addition and the set of edge functions containing only the identity function and the zero function.

The rest of this section assumes a fixed flow domain $(M, +, 0, E)$ and a (potentially infinite) set of nodes $\mathfrak{N}$.

Definition 7.3   Graph.   A *(partial) graph* $G = (N, e)$ consists of a finite set of nodes $N \subseteq \mathfrak{N}$ and a mapping from pairs of nodes to edge functions $e: N \times \mathfrak{N} \to E$.

A *flow of graph* $G = (N, e)$ *under inflow in*: $N \to M$ is a solution of the following fixpoint equation (the same as (FlowEqn), repeated for clarity) over G, denoted FlowEqn(in, e, fl):

$$\forall n \in \mathrm{dom}(in).\ fl(n) = in(n) + \sum_{n' \in \mathrm{dom}(in)} e(n', n)(fl(n'))$$

Note that every graph need not have a flow; there are graphs (N, e) and inflows *in* for which there does not exist a solution fl to FlowEqn(in, e, fl). On the other hand, graphs that have a flow are called flow graphs.

Definition 7.4   Flow Graph.   A *flow graph* $H = (N, e, fl)$ consists of a graph (N, e) and a function $fl: N \to M$ such that there exists an *inflow in*: $N \to M$ satisfying FlowEqn(in, e, fl).

We let $\mathrm{dom}(H) = N$, and sometimes identify H and $\mathrm{dom}(H)$ to ease notational burden. Two flow graphs with disjoint domains always compose to a graph, but this will only be a flow graph if their flows are chosen consistently to admit a solution to the resulting flow equation (thus, the flow graph algebra below has a special *invalid* element $H_\sharp$, which is the result of flow graph composition of incompatible flow graphs).

Definition 7.5   Flow Graph Algebra.   The *flow graph algebra* (FG, $\odot$, $H_\emptyset$) for flow domain $(M, +, 0, E)$ is defined by

$$\mathrm{FG} ::= H \in \big\{(N, e, fl) \mid (N, e, fl) \text{ is a flow graph}\big\} \mid H_\sharp$$

$$(N_1, e_1, fl_1) \odot (N_2, e_2, fl_2) := \begin{cases} H & H = (N_1 \uplus N_2, e_1 \uplus e_2, fl_1 \uplus fl_2) \in \mathrm{FG} \\ H_\sharp & \text{otherwise} \end{cases}$$

$$\_ \odot H_\sharp := H_\sharp \odot \_ := H_\sharp$$

$$H_\emptyset := (e_\emptyset, fl_\emptyset)$$

where $e_\emptyset$ and $fl_\emptyset$ are the edge functions and flow on the empty set of nodes $N = \emptyset$.

Cancellativity of the flow domain operator $+$ is key to defining an abstraction of flow graphs that permits local reasoning. The following lemma follows from the fact that $+$ is cancellative.

Lemma 7.6   *Given a flow graph* $(N, e, fl) \in \mathrm{FG}$, *there exists a unique inflow in*: $N \to M$ *such that* FlowEqn(in, e, fl).

Proof. Suppose in and in' are two solutions to $\mathsf{FlowEqn}(\_, e, \mathit{fl})$. Then, for any n,

$$\mathit{fl}(n) = in(n) + \sum_{n' \in \mathrm{dom}(in)} e(n', n)(\mathit{fl}(n')) = in'(n) + \sum_{n' \in \mathrm{dom}(in')} e(n', n)(\mathit{fl}(n'))$$

which, by cancellativity of the flow domain, implies that $in(n) = in'(n)$. $\qquad\square$

Our abstraction of flow graphs consists of two complementary notions. Lemma 7.6 implies that any flow graph has a unique inflow. Thus, we can define an inflow function that maps each flow graph $H = (N, e, \mathit{fl})$ to the unique inflow $\mathsf{inf}(H) \colon H \to M$ such that $\mathsf{FlowEqn}(\mathsf{inf}(H), e, \mathit{fl})$. We can also define the *outflow* of H as the function $\mathsf{outf}(H) \colon \mathfrak{N} \setminus N \to M$ defined by

$$\mathsf{outf}(H)(n) := \sum_{n' \in N} e(n', n)(\mathit{fl}(n')).$$

Definition 7.7   Flow Interface.   Given a flow graph $H \in \mathsf{FG}$, its *flow interface* $\mathsf{int}(H)$ is the tuple $(\mathsf{inf}(H), \mathsf{outf}(H))$ consisting of its inflow and its outflow.

We use $I.in$ and $I.out$ to denote, respectively, the inflow and outflow of an interface I. We can finally define the flow interface resource algebra, which will allow us to use flow interfaces in our Iris proofs.

Definition 7.8   Flow Interface Algebra.   The flow interface algebra $(\mathsf{FI}, \mathcal{V}, \oplus, I_\emptyset)$ is defined as

$$\mathsf{FI} ::= I \in \{\mathsf{int}(H) \mid H \in \mathsf{FG}\} \mid I_{\frac{1}{4}} \qquad \mathcal{V}(a) := a \neq I_{\frac{1}{4}} \qquad I_\emptyset := \mathsf{int}(H_\emptyset)$$

$$I_1 \oplus I_2 := \begin{cases} \mathsf{int}(H) & \exists H_1\, H_2.\, H = H_1 \odot H_2 \wedge \forall i \in \{1, 2\}.\, \mathsf{int}(H_i) = I_i \\ I_{\frac{1}{4}} & \text{otherwise.} \end{cases}$$

Theorem 7.9   *The flow interface algebra* $(\mathsf{FI}, \mathcal{V}, \oplus, I_\emptyset)$ *is a resource algebra.*

The proof of this Theorem is included in the machine-checked proofs accompanying this monograph (see Chapter 13). The proof follows from the definitions of flow interfaces and RAs. It is worth noting that cancellativity of flow domains is crucial to proving that composition of flow interfaces is a well-defined function. We can now use flow interfaces as ghost values in Iris proofs. In fact, we use them extensively in this monograph in order to reason about global graph quantities such as keysets.

7.4   ENCODING KEYSETS USING FLOWS

We now answer the question of how to define a keyset function in a node-local way, using flow interfaces.

To define keysets using flows, we build on the concept of edgesets. Recall that the edgeset $\mathsf{es}(n, n')$ is the set of keys for which an operation arriving at a node n traverses (n, n'). This is a node-local concept, and hence can be expressed in Iris. Let the *inset* of a node n, written $\mathsf{ins}(n)$, be defined by the following fixpoint equation

$$\forall n \in N. \; \mathsf{ins}(n) = in(n) \cup \bigcup_{n' \in N} \mathsf{es}(n', n) \cap \mathsf{ins}(n')$$

where $in(n) := (n = r \; ? \; \mathbb{K} : \emptyset)$. The inset of a node n is thus $\mathbb{K}$ if n equals the root r, else the set of keys k that are in the inset of a predecessor n' such that $k \in \mathsf{es}(n', n)$. Intuitively, $\mathsf{ins}(n)$ is the set of keys for which operations could potentially arrive at n in a sequential setting. For example, in Figure 6.2 insets are shown in the top-left of each node; $\mathsf{ins}(y_2) = [5, 8)$ and $\mathsf{ins}(n') = [5, \infty)$. Let the *outset* of n, $\mathsf{outs}(n)$, be the keys in the union of edgesets of edges leaving n. The keyset can then be defined as $\mathsf{ks}(n) = \mathsf{ins}(n) \setminus \mathsf{outs}(n)$.

If the equation defining the inset looks familiar, the reason is that it is just (FlowEqn) with ins in place of fl, using sets and set operations, and edge functions that take the intersection with the appropriate edgeset. This means we can define a flow domain where the flow at each node is the inset of that node. This will allow us to talk about the keyset in terms of node-local conditions: in particular, we can now use the keyset in the keyset RA ghost state described in §5.2.

Encoding the inset as a flow requires using multisets of keys[3] as the flow domain. We label each edge (n, n') in a graph G by the function $e_{\mathsf{es}(n,n')} := (\lambda X. \, \mathsf{es}(n, n') \cap X)$. If the global inflow is $in = (\lambda n. \; (n = r \; ? \; \mathbb{K} : \emptyset))$, which encodes the fact that operations on all keys k start at the root r, then the flow equation implies that $\mathit{fl}(n)$ is the inset of n.

As with the keyset RA, we use an authoritative version of the flow interface RA (AUTH(FI), see Definition 5.2) in our proofs. This enables us to reason about the concurrent setting where threads can lock nodes and take them out of the shared state. Since we are now using two kinds of ghost state, we use γ_I to denote the AUTH(FI) ghost location and γ_k for the AUTH(KEYSET) ghost location.

The resulting definitions are shown in Figure 7.2. The inset, outset, and keyset are defined in terms of a flow interface as described above (we overload the same symbols used before because they express the same quantities). The node predicate $\mathsf{N}(n, C_n)$ now has a fragment of both the keyset RA and flow interface RA corresponding to the node n. Note that n's keyset is now a function of n's singleton interface and n, making it local as desired. The predicate φ contains the constraints on the interface of the global search structure which are needed to make the flow

---

[3]We cannot use sets of keys because a flow domain is a cancellative commutative monoid (Definition 7.1), and unlike multiset union, set union is not cancellative.

$$\mathsf{ins}(I, n) := I.in(n)$$

$$\mathsf{outs}(I, n') := I.out(n') \qquad \mathsf{outs}(I) := \bigcup_{n' \notin \mathsf{dom}(I)} \mathsf{outs}(I, n')$$

$$\mathsf{ks}(I, n) := \mathsf{ins}(I, n) \setminus \mathsf{outs}(I, n)$$

$$\mathsf{N}(n, I_n, C_n) := \mathsf{Node}(n, I_n, C_n) * \lceil \circ(\mathsf{ks}(I_n, n), C_n) \rceil^{\gamma_k} * \lceil \circ I_n \rceil^{\gamma_I} * \mathsf{dom}(I_n) = \{n\}$$

$$\varphi(r, I) := \mathcal{V}(I) \wedge I.in(r) = \mathbb{K} \wedge I.out = \lambda_0$$

$$\mathsf{CSS}(r, C) := \exists I. \lceil \bullet I \rceil^{\gamma_I} * \varphi(r, I) * \lceil \bullet(\mathbb{K}, C) \rceil^{\gamma_k}$$
$$* \mathop{\text{\Large *}}_{n \in \mathsf{dom}(I)} \Big(\exists b\, C_n.\ \mathsf{L}(b, n, \mathsf{N}(n, C_n)) \Big)$$

Figure 7.2: The definition of CSS and other predicates using a flow-based encoding of keysets.

of each node be interpreted as the inset of each node. CSS contains the authoritative global interface $\lceil \bullet I \rceil^{\gamma_I}$, as well as the authoritative keyset RA element as before.

With such a definition, we can reason about search structures with an unbounded number of nodes. In the next chapter we will see how to prove three templates for real-world search structures.

CHAPTER 8

Verifying Single-Copy Concurrent Search Structures

This chapter shows how to bring together all the concepts developed so far in order to verify template algorithms for single copy concurrent search structures (in which each key is present at most once in the given search structure). The three templates we describe are based on three common concurrency techniques: give-up, link, and lock-coupling.

$$\langle\, C.\ \mathsf{CSS}(r, C)\,\rangle\ \mathtt{cssOp}\ \omega\ k\ \langle\, res.\ \mathsf{CSS}(r, C') * \Psi_\omega(k, C, C', res)\,\rangle$$

$$\Psi_\omega(k, C, C', res) := \begin{cases} C' = C \wedge (res \iff k \in C) & \omega = \mathtt{search} \\ C' = C \cup \{k\} \wedge (res \iff k \notin C) & \omega = \mathtt{insert} \\ C' = C \setminus \{k\} \wedge (res \iff k \in C) & \omega = \mathtt{delete} \end{cases}$$

Figure 8.1: The atomic specification of core search structure operations. We prove that all single-copy search structure templates in this monograph satisfy this specification.

Remember, our aim is to prove the atomic specification shown in Figure 8.1 for the template method cssOp that represents, via the parameter ω, an arbitrary search structure operation (either search, insert, or delete). This specification uses an abstract predicate $\mathsf{CSS}(r, C)$ that represents a search structure with root r containing the set of keys C. The binder on C in the precondition is a special pseudo-quantifier (see Chapter 3) that captures the fact that during the execution of ω, the value of C can change (e.g., by concurrent operations) but at the linearization point, ω on operation key k changes $\mathsf{CSS}(r, C)$ to $\mathsf{CSS}(r, C')$ in an atomic step. The new set of keys C', and the eventual return value res, satisfy the predicate $\Psi_\omega(k, C, C', res)$—here C is bound to the contents *just before* the linearization point. The bottom line is that clients of the search structure can pretend that they are using a sequential implementation with specification Ψ_ω.

$$\mathsf{InFP}(n) \twoheadrightarrow \big\langle C. \ \mathsf{CSS}(r, C) \big\rangle \ \texttt{lockNode} \ n \ \big\langle \mathsf{CSS}(r, C) * \mathsf{N}(n, I_n, C_n) \big\rangle$$

$$\mathsf{N}(n, I_n, C_n) \twoheadrightarrow \big\langle C. \ \mathsf{CSS}(r, C) \big\rangle \ \texttt{unlockNode} \ n \ \big\langle \mathsf{CSS}(r, C) \big\rangle$$

Figure 8.2: High-level specifications for the lock module used by all template proofs in this chapter. These can be proved from the low-level specifications we proved in Chapter 2 and the definitions of CSS and InFP.

All the template proofs in this chapter assume some implementation of `lockNode` and `unlockNode` that satisfy the abstract specifications from Figure 4.6.[1] To simplify the upcoming proofs, we will assume that `lockNode` and `unlockNode` satisfy the higher-level specifications shown in Figure 8.2, which can be proved easily from the above specifications and the definitions of CSS and InFP.

These specifications are written in a new form, using the magic wand $\twoheadrightarrow$, which is the ownership analog of implication (see Chapter 3 for more details). The specification of `lockNode` says that if we own $\mathsf{InFP}(n)$, i.e., if we know that n is in the footprint of the structure, then we can use the atomic triple that follows. The triple tells us that `lockNode` operates on the shared state of the search structure $\mathsf{CSS}(r, C)$, and transfers ownership of the node-local resources $\mathsf{N}(n, I_n, C_n)$ from the shared state to the caller's local state while returning the search structure unmodified.

Similarly, the specification of `unlockNode` says that if a thread owns the node predicate for n, then it can call `unlockNode` in an atomic step to put n back into the shared search structure.

8.1   THE GIVE-UP TEMPLATE

We start by considering a template algorithm that uses the well-known *give-up* style of concurrency (Figure 8.3). The proof of this template is very similar to that of the link template, but simpler, so we describe it first.

The give-up template, like the link template, uses locks only when reading or writing from a node and does not hold locks while traversing from one node to the next. Unlike the link template, there are no link edges added by threads that move data from one node to another. Instead, each node stores a *range* field: this is an under-approximation of that node's inset.[2] Upon arriving at a new node n, each thread locks the node and checks its operation key k against the range of n. If k is in the range of n then the thread knows that it is still on the correct path, and it continues. If not, it gives up: it relinquishes the lock on n and goes back to the root of the data

---

[1]We showed in Chapter 4 that a simple spin-lock implementation satisfies these specifications. However, note that one can use more complex lock implementations, as long as they satisfy these specifications.

[2]In practice, the range of a node is nearly always equal to its inset, except for brief moments in between some maintenance operations such as splits.

```
1 let create () =                        15 let rec cssOp ω r k =
2   let r = allocRoot () in               16   let n = traverse r r k in
3   r                                     17   match decisiveOp ω n k with
4                                         18   | None ->
5 let rec traverse r n k =                19     unlockNode n;
6   lockNode n;                           20     cssOp ω r k
7   if inRange n k then                   21   | Some res ->
8     match findNext n k with             22     unlockNode n;
9     | None -> n                         23     res
10    | Some n' -> unlockNode n;
11       traverse n' k
12  else
13    unlockNode n;
14    traverse r r k
```

Figure 8.3: The give-up template algorithm. The cssOp method is the main method, and represents all the core search structure operations via the parameter ω. It makes use of an auxiliary method traverse that recursively traverses the search structure until it finds the node upon which to operate.

structure to retry. (In fact, it could go back to any previous node to retry, but eventually it might have to go back to the root. For simplicity, we consider the version of the algorithm that goes back to the root immediately.)

The code for this template algorithm is given in Figure 8.3. Note that in addition to the helper functions findNext and decisiveOp, this template assumes a helper function inRange. When called as inRange $n\,k$, this function returns *true* if and only if k is in the range of n.

The give-up template can be instantiated by a B+ tree, for instance, by adding two additional fields to each node n. These fields keep track of lower and upper bounds for keys that are present in the subtree rooted at n. When a thread looking for k arrives at a node n, it checks whether k is in the range of n (by checking whether k is between the lower and upper bounds) or not. The thread gives up and restarts if not. Though we have conceived a range as consisting of a lower and upper bound, in fact a range can be an arbitrary function as long as it is a subset of the inset. For example, it can be a set of key values that hash to a particular value for a hash table.

8.1.1   PROOF OF THE GIVE-UP TEMPLATE

The definition of the search structure predicate CSS for the give-up template is given in Figure 8.4. This is almost the same as the one we developed in the last chapter using flows (see Figure 7.2 in §7.4), but with an extra form of ghost state that is used to reason about traversals. The reason we need additional ghost state is that the give-up template performs a traversal over the search structure where it holds no locks when moving from one node to the next. For instance,

$$\text{ins}(I, n) := I.in(n)$$

$$\text{outs}(I, n') := I.out(n') \qquad \text{outs}(I) := \bigcup_{n' \notin \text{dom}(I)} \text{outs}(I, n')$$

$$\text{ks}(I, n) := \text{ins}(I, n) \setminus \text{outs}(I, n)$$

$$\text{N}(n, I_n, C_n) := \text{Node}(n, I_n, C_n) * \lceil \circ(\text{ks}(I_n, n), C_n) \rceil^{\gamma_k} * \lceil \circ I_n \rceil^{\gamma_I} * \text{dom}(I_n) = \{n\}$$

$$\varphi(r, I) := \mathcal{V}(I) \wedge I.in(r) = \mathbb{K} \wedge I.out = \lambda_0$$

$$\text{CSS}(r, C) := \exists I. \lceil \bullet I \rceil^{\gamma_I} * \varphi(r, I) * \lceil \bullet(\mathbb{K}, C) \rceil^{\gamma_k} * \lceil \bullet \text{dom}(I) \rceil^{\gamma_f}$$
$$* \underset{n \in \text{dom}(I)}{\LARGE *} \left(\exists b \, I_n \, C_n . \, \text{L}(b, n, \text{N}(n, I_n, C_n)) \right)$$

Figure 8.4: The definition of CSS and other predicates used by the give-up template proof.

at the point when the `traverse` method is called, the node n is not locked. Yet, to be able to apply the specification of `lockNode`, we need to know that n is a node in the data structure (and not some arbitrary memory address that may not be allocated).

We solved this issue in the two-node template proof by defining a predicate $\text{InFP}(n_1, n_2, n)$, that asserted that n is one of the two nodes $\{n_1, n_2\}$ in the structure. This approach no longer works since we are in the general case where the structure has an unbounded number of nodes that are not known to us a priori.

Our solution is to use an authoritative RA of sets of nodes, $\text{AUTH}(\mathfrak{N})$, at a new ghost location γ_f. The authoritative element contained in CSS is $\lceil \bullet \text{dom}(I) \rceil^{\gamma_f}$, and this captures the domain of the shared state (which is equal to the domain of the global flow interface). The following properties of $\text{AUTH}(\mathfrak{N})$ allow threads to take snapshots of the footprint and assert locally that a given node is in the footprint:

$$\frac{\text{AUTH-SET-UPD}}{X \subseteq Y} \qquad\qquad \frac{\text{AUTH-SET-SNAP}}{\bullet X \rightsquigarrow \bullet X \cdot \circ X} \qquad\qquad \frac{\text{AUTH-SET-VALID}}{\mathcal{V}(\bullet X \cdot \circ Y)}$$

We then define the footprint predicate as

$$\text{InFP}(n) := \lceil \circ \{n\} \rceil^{\gamma_f}$$

$\{\mathsf{Node}(n, I_n, C_n)\}$
`inRange` n k
$\{v. \mathsf{Node}(n, I_n, C_n) * (v = \mathit{true} \Rightarrow k \in \mathsf{ins}(I_n, n))\}$

$\{\mathsf{Node}(n, I_n, C_n) * k \in \mathsf{ins}(I_n, n)\}$
`findNext` n k
$$\left\{ \begin{array}{l} v. \mathsf{Node}(n, I_n, C_n) * \Big(v = \mathsf{None} * k \notin \mathsf{outs}(I_n) \\ \qquad\qquad\qquad \lor v = \mathsf{Some}(n') * k \in \mathsf{outs}(I_n, n') \Big) \end{array} \right\}$$

$\{\mathsf{Node}(n, I_n, C_n) * k \in \mathsf{ks}(I_n, n)\}$
`decisiveOp` ω n k
$$\left\{ \begin{array}{l} v. \mathsf{Node}(n, I_n, C_n') * \Big(v = \mathsf{None} * C_n = C_n' \\ \qquad\qquad\qquad \lor v = \mathsf{Some}(v') * \Psi_\omega(k, C_n, C_n', v') \Big) \end{array} \right\}$$

$\mathsf{Node}(n, I_n, C_n) * \mathsf{Node}(n, I_n', C_n') \mathrel{-\!\!*} \mathsf{False}$

Figure 8.5: The assumptions made by the give-up template on implementations.

which expresses ownership of a fragment $\{n\}$ of the domain. By AUTH-SET-VALID, this along with $\boxed{\bullet\,\mathsf{dom}(I)}^{\gamma_f}$ implies that $n \in \mathsf{dom}(I)$. Threads can thus use AUTH-SET-SNAP and AUTH-FRAG-OP to create the resource $\mathsf{InFP}(n)$, which can then be moved into the thread's local state.

Before we move on to the proof of the give-up template, let us review the assumptions made by the give-up template on implementations (Figure 8.5). As usual, they are all Hoare triples that operate on the abstract Node predicate, meaning their proofs need not reason about concurrency. We require that $\mathtt{inRange}\, n\, k$ return a Boolean value, which if true implies that k is in the inset of n. (If it is false, we do not require any additional information, because the algorithm gives up and restarts from the root.) The $\mathtt{findNext}$ method is used by the traversal at each step: given a node n and a key k, it either returns None, indicating that there is no outgoing edge with k in its edgeset, or returns $\mathsf{Some}(n')$ such that k is in the edgeset of (n, n'). Neither of these two methods modify the given node n. The $\mathtt{decisiveOp}$ method has a similar spec to what we have used before, except that now we allow it to also fail: if it returns None then we require it to return the node n with unmodified contents, while if it returns $\mathsf{Some}(v')$ then we require it to satisfy the per-operation specification $\Psi_\omega(k, C_n, C_n', v')$ as before. Finally, we assume that the heap representation predicate $\mathsf{Node}(n, I_n, C_n)$ implies that we have ownership of the heap location n; in particular, we need the property that it cannot be duplicated, hence owning two copies of it implies False.

```
1 InFP(n) -*⟨C. CSS(r, C)⟩
2 let rec traverse r n k =
3   {InFP(n)}
4   lockNode n;
5   {N(n, In, Cn)}
6   if inRange n k then
7     {N(n, In, Cn) * k ∈ ins(In, n)}
8     match findNext n k with
9     | None -> {N(n, In, Cn) * k ∈ ins(In, n) * k ∉ outs(In)}
10       n
11    | Some n' -> {N(n, In, Cn) * k ∈ ins(In, n) * k ∈ outs(In, n')}
12      {N(n, In, Cn) * InFP(n')}
13      unlockNode n; {InFP(n')}
14      traverse n' k
15  else
16    {N(n, In, Cn)}
17    unlockNode n;
18    {InFP(r)}
19    traverse r r k
20 ⟨v. CSS(r, C) * N(v, Iv, Cv) * k ∈ ks(Iv, v)⟩
```

Figure 8.6: The proof of the `traverse` method of the give-up template algorithm.

Moving on to the proofs, we prove the following specification for the `traverse` helper function:

$$\mathsf{InFP}(n) \; -\!\!* \; \big\langle C. \; \mathsf{CSS}(r, C)\big\rangle \; \texttt{traverse} \; r \; n \; k \; \big\langle v. \; \mathsf{CSS}(r, C) * \mathsf{N}(v, I_v, C_v) * k \in \mathsf{ks}(I_v, v)\big\rangle$$

This specification says that if n is in the footprint of the search structure, then `traverse` operates on the entire search structure and returns a locked node v such that k is in the keyset of v, while leaving the global contents unmodified.

The proof of `traverse` is shown in Figure 8.6. Note that the $\mathsf{InFP}(n)$ predicate from the precondition is available in our proof context at the beginning of the proof. Hence, we can use it to apply the high-level `lockNode` specification (Figure 8.2) to get the node $\mathsf{N}(n, I_n, C_n)$ into the thread-local state. We then use the specification of `inRange` from Figure 8.5.

In the then branch, this gives us the additional predicate $k \in \mathsf{ins}(I_n, n)$. We use this to apply the specification of `findNext`, which leads to two cases. In the case where `findNext` returns None, the specification tells us that we have $k \in \mathsf{ins}(I_n, n) * k \notin \mathsf{outs}(I_n)$. By the definition of ks in Figure 8.4, this is exactly $k \in \mathsf{ks}(I_n, n)$, which allows us to prove the postcondition. (Technically, this is the linearization point, and we use the appropriate proof rules to "commit" the change to the shared structure and establish the postcondition.)

```
 1 ⟨C. CSS(r, C)⟩
 2 let rec cssOp ω r k =
 3   {InFP(r)}
 4   let n = traverse r r k in
 5   {N(n, I_n, C_n) * k ∈ ks(I_n, k)}
 6   match decisiveOp ω n k with
 7   | None -> {N(n, I_n, C_n)}
 8     unlockNode n; {True}
 9     cssOp ω r k
10   | Some res ->
11     {Node(n, I_n, C'_n) * Ψ_ω(k, C_n, C'_n, res) * k ∈ ks(I_n, n) * ···}
12     (* Linearization point: open precondition *)
13     ⟨⌜○(ks(I_n, n), C_n)⌝^{γ_k} * Ψ_ω(k, C_n, C'_n, res) * k ∈ ks(I_n, n) * ⌜•(𝕂, C)⌝^{γ_k} * ···⟩
14     ⟨⌜○(ks(I_n, n), C'_n)⌝^{γ_k} * Ψ_ω(k, C, C', res) * ⌜•(𝕂, C')⌝^{γ_k} * ···⟩  (* By KS-UPD *)
15     ⟨N(n, I_n, C'_n) * CSS(r, C') * Ψ_ω(k, C, C', res)⟩
16     unlockNode n;
17     ⟨CSS(r, C') * Ψ_ω(k, C, C', res)⟩  (* Prove postcondition, commit *)
18     {True}
19     res
20 ⟨v. CSS(r, C') * Ψ_ω(k, C, C', v)⟩
```

Figure 8.7: The proof of the give-up template algorithm.

If findNext returns Some(n'), then we have $k \in \mathsf{ins}(I_n, n) * k \in \mathsf{outs}(I_n, n')$ from findNext's postcondition. Before we move on to the unlocking and give away the node predicate, note that we need to prove $\mathsf{InFP}(n')$ in order to satisfy the precondition of the recursive call to traverse. We obtain this by opening the precondition and using $k \in \mathsf{outs}(I_n, n')$ along with $\varphi(I)$. Since I_n is a sub-interface of I, and $\varphi(I)$ specifies that I has no outflow, n can have an outflow of k to n' only if n' is also a node in I. We thus obtain $\mathsf{InFP}(n')$, and this completes this branch.

In the else branch of the call to inRange, we simply use the high-level specification of unlockNode to give back the node predicate. We can then use induction to assume that the recursive call to traverse will give us the desired postcondition. The precondition of the recursive call requires us to prove $\mathsf{InFP}(r)$, but note that this is always true from the definition of CSS (as $\varphi(r, I) \Rightarrow r \in \mathsf{dom}(I)$). This completes the proof of traverse.

We can now verify cssOp. As mentioned above, the definition of CSS gives us $\mathsf{InFP}(r)$, so we open the precondition to get $\mathsf{InFP}(r)$ and use it to apply the specification of traverse. This gives us a state where we have a node n such that k is in the keyset of n, which is all we need to apply the specification of decisiveOp. In the case where decisiveOp fails and returns

None, we simply use the high-level specification of `unlockNode` to give back the node n and retry (using induction to handle the recursive call).

On the other hand, if `decisiveOp` succeeds, then we have the state shown in line 11 (we write $\cdots$ to hide the resources irrelevant to the next step in the proof). Note that we cannot yet write $N(n, I_n, C'_n)$ because `decisiveOp` modified only Node and not the rest of the ghost state in N. We update this ghost state as well as the ghost state in the shared state in the next step. Since the call to `unlockNode` is the linearization point, we open the precondition and prepare to commit our changes to the shared state. In particular, we focus on the keyset RA ghost state, as shown in line 13. We now have all the resources needed to use the rule KS-UPD and update both the contents of n in its fragment from C_n to C'_n and the global contents from C to some C' such that $\Psi_\omega(k, C, C', res)$ holds. The state that we have now has all the ghost state updated appropriately, so we can fold the node n as $N(n, I_n, C'_n)$. We can then use the specification of `unlockNode` to give the node n back to the shared state, and prove the postcondition of `cssOp` in order to finish the proof.

8.1.2 MAINTENANCE OPERATIONS

The template algorithms we have seen so far cover the core search structure operations search, insert, and delete. However, real search structures also need maintenance operations in order to function correctly. For example, in a B+ tree, successive inserts of nearby keys can make a leaf node full, necessitating a split operation to split the leaf into two. Conversely, practical B+-tree implementations also have a free-at-empty operation when a node becomes empty. (Classically, B+-trees use "merges" when neighboring nodes become less than half-full. However, free-at-empty turns out to be more efficient, because merged nodes tend to be split soon afterward if, as is common, there are more inserts than deletes to the structure.) To ensure that these maintenance operations do not invalidate any core operations, we must verify that they (a) preserve the invariants of the search structure, and (b) do not change the contents of the entire structure.

Both these conditions can be expressed in the following specification for an arbitrary maintenance operation `maintOp`:

$$\langle C.\ \text{CSS}(r, C) \rangle\ \texttt{maintOp}\ r\ \langle \text{CSS}(r, C) \rangle$$

This triple gives the maintenance operation the permission to modify the search structure $\text{CSS}(r, C)$ as long as it appears, to any other thread, to instantaneously modify the structure to another valid search structure $\text{CSS}(r, C)$ with the same contents. If the maintenance operation operates on locked nodes, then its proof can again be split between a concurrent proof of the `maintOp` and a sequential proof of the operations it performs on locked nodes. We omit these proofs here, but they can be proved using the same techniques that we have shown so far.

For the give-up template, preserving CSS includes preserving the invariant that the range of every node is a subset of its inset. In the B+ tree example, the range of the root at any given time is the entire key space $\mathbb{K}$. The range of a non-root node n is set to its inset when n is created.

If n is later split or merged, then its range is reduced to the inset it will be at the end of the split or merge. A thread searching for k might have visited the parent of n before the split began and then might visit node n after n was split. The thread would then see that the key k is not in the range of n and give up.

The split operation involves creating a new node, and to show that this does not invalidate any of the flow-based invariants, we use the following notion.[3]

Definition 8.1   A flow interface I' is a *domain extension* of interface I, written $I \Subset I'$, if and only if

1. $\mathsf{dom}(I) \subseteq \mathsf{dom}(I')$,

2. $\forall n \in I.\ I.in(n) = I'.in(n)$, and

3. $\forall n' \notin I'.\ I.out(n') = I'.out(n')$.

This definition allows I' to differ from I by having a larger domain, as long as the new nodes are fresh and edges from the new nodes do not change the outflow of the modified region. We can show that replacing an interface with a domain extension is a frame-preserving update in the flow interface RA:

FLOWINT-DOM-UPD
$$\frac{I_1 \Subset I_1' \qquad I_1' \cap I_2 = \emptyset \qquad \forall n \in I_1' \setminus I_1.\ I_2.out(n) = 0}{(\bullet I_1 \oplus I_2, \circ I_1) \rightsquigarrow \left\{ (\bullet I_1' \oplus I_2, \circ I_1') \mid (I_1 \oplus I_2) \Subset (I_1' \oplus I_2) \right\}}$$

We use this lemma in the proof of the split operation to update the flow interface ghost state when adding the new node.

8.2   THE LINK TEMPLATE

The link template is very similar to the give-up template, and shares much of its proof. While the link template algorithm might even look simpler than the give-up, for it does not have the `inRange` helper function, it is in fact a bit more complicated to prove. This section explains why, and then shows how to extend the proof of the give-up template to the link template.

8.2.1   INREACH

The main reason the give-up proof does not work for the link template is that without `inRange`, it is hard to prove that the link template's traversal stays on track and finds the correct node (i.e., a node n with $k \in \mathsf{ks}(n)$ at the end.

In the absence of concurrent operations (particularly concurrent split operations), this follows because we start off at the root r, where by definition $k \in \mathsf{ins}(r)$, and traverse an edge

---

[3]Remember, we write I for $\mathsf{dom}(I) = \mathsf{dom}(I.in)$ when it is clear from context.

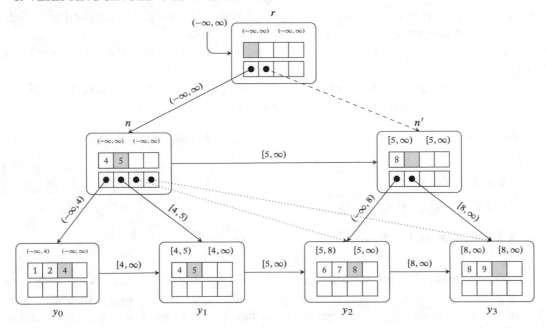

Figure 8.8: An example B-link tree in the middle of a split. Node n was full, and has been half-split and children y_2 and y_3 have been transferred to the new node n' (old edges are shown with dotted lines), but the complete-split has yet to add n' to the parent r (the dashed edge). Each node now additionally contains its inset (see §7.4) in the top left and the inreach (defined in this section) in the top right. The label with a curved arrow to the top-left of the root is its inflow (explained in §7.4).

(n, n') only when $k \in \mathsf{es}(n, n')$, maintaining the invariant that $k \in \mathsf{ins}(n)$. When a node n has no outgoing edge having k in its edgeset, we know by definition that $k \in \mathsf{ks}(n)$.

In the presence of concurrent split operations, the $k \in \mathsf{ins}(n)$ invariant no longer holds because the inset of a node n shrinks after a split. For example, Figure 8.8 shows a B-link tree state in between the half-split and full-split of n. When the full-split completes and r is linked to n' (Figure 8.9), then the inset of n will be reduced from $(-\infty, \infty)$ to $(-\infty, 5)$ as all keys larger than 5 will go from r directly to n'. This means that an operation looking for a key $k > 5$ which was at n before the split will now find itself at a node such that $k \notin \mathsf{ins}(n)$.

Fortunately, the operation is not lost: if it traverses the link edge, it will arrive at a node with k in its inset (namely, n'). This means that if we add k back to the inset of n, then we would not be changing the keyset of any node: k will not be in n's keyset as it is in the edgeset of the link edge, and k is already in the inset of n'. Because this quantity is no longer the inset (as k would not arrive at n in a sequential setting), we call this the *inreach* of n, written $\mathsf{inr}(n)$ (intuitively, this is the set of keys k that can start at n and, following edges having k in their

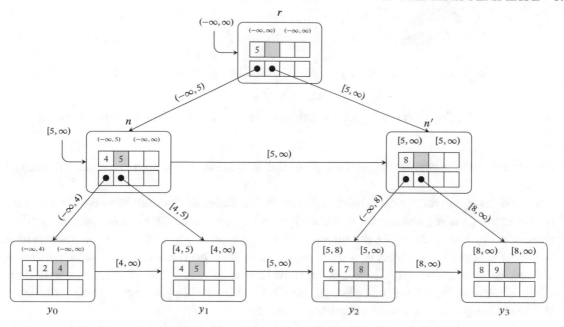

Figure 8.9: The B-link tree from Figure 8.8 after the full-split of n has completed. Note that while n's inset has been reduced to $(-\infty, 5)$, its inreach is preserved at $(-\infty, \infty)$.

edgeset, reach the node containing k in its keysct). Figure 8.8 shows the inreach of each node in its top-right corner. For example, the inreach of y_2 is $[5, \infty)$ despite its inset's being only $[5, 8)$ because it can still reach nodes with keys in $[8, \infty)$ in their keyset via link edges.

We define the inreach to be the solution to the following fixpoint equation

$$\forall n \in N. \; \mathsf{inr}(n) = in(n) \cup \bigcup_{n' \in N} \mathsf{es}(n', n) \cap \mathsf{inr}(n')$$

where in is any inflow such that $in(r) = \mathbb{K}$. This may look identical to the definition of inset, but there is a subtle, but vital, difference: by not constraining the inflow of non-root nodes, we enable the split operation to add flow to nodes it has split to ensure that their inreach records the fact that they can still reach keys k that were moved to other nodes. For example, in Figure 8.9 when the full-split adds the edge (r, n') and re-routes keys in $[5, \infty)$ to take (r, n') instead of (r, n), then n's inset is reduced from $(-\infty, \infty)$ to $(-\infty, 5)$.

Our solution to this is to *increase the inflow* of n to make up for the removed keys. We will formally justify this operation in §8.2.3, but at a high level what we do is increase $in(n)$ from $\emptyset$ to $[5, \infty)$, thereby preserving its inreach of $(-\infty, \infty)$. Intuitively, this operation does not violate any other invariant because the newly added keys $[5, \infty)$ are propagated via the link edge to a

$$\mathsf{inr}(I_n, n) := I_n.in(n) \qquad\qquad \mathsf{outs}(I_n) := \bigcup_{n' \notin \mathsf{dom}(I_n)} \mathsf{outs}(I_n, n')$$

$$\mathsf{outs}(I_n, n') := I_n.out(n') \qquad\qquad \mathsf{ks}(I_n, n) := \mathsf{inr}(I_n, n) \setminus \mathsf{outs}(I_n, n)$$

Figure 8.10: The definition of keyset and associated quantities for the link template proof.

node that has those keys in its inset (n'). Thus, this increase in inflow does not change the keyset of any node.

We have one final issue to solve: as it stands, the full-split does not preserve the interface of the region $\{r, n, n'\}$ because the outflow to y_2 and y_3 has increased. For example, the outflow from n' to y_2 was $[5, 8)$ before the full-split (the edge label in Figure 8.8 is the edgeset, which is $(-\infty, 8)$, but since the inset of n' is $[5, \infty)$, keys below 5 do not arrive at n' to be part of its inflow). However, after the full-split, the outflow from n' to y_2 is $\{[5, 8), [5, 8)\}$, i.e., the multiset where all keys in $[5, 8)$ have multiplicity 2. This is because one copy of keys in $[5, 8)$ arrive from the newly introduced inflow on n, and one copy from r.

The problem is in our encoding of edgesets and inreach in the flow framework: we had to use multisets instead of sets because sets are not a flow domain. Our solution is to tweak the edge functions from $e_{\mathsf{es}(n,n')} := (\lambda X. \, \mathsf{es}(n, n') \cap X)$ to $e_{\mathsf{es}(n,n')} := (\lambda X. \, \{k \rightarrowtail (k \in \mathsf{es}(n, n') \cap X \, ? \, 1 : 0)\})$, essentially projecting the multiset intersection back to a set, and preventing multiple copies of keys from being propagated.

We now have an invariant for `traverse`: $k \in \mathsf{inr}(n)$. This is true at the root, because $\mathbb{K} = in(r) \subseteq \mathsf{inr}(r)$, and it is preserved during traversal since `findNext` follows edges with k in the edgeset. We will ensure that no concurrent operations reduce the inreach of any node by adding an appropriate constraint to the search structure predicate CSS in §8.2.2.

We amend the definition of keyset used in the give-up proof accordingly (Figure 8.10). The inreach is defined to be the flow of each node, the definition of outset is unchanged and the keyset of each node n is defined to be $\mathsf{inr}(n) \setminus \mathsf{outs}(n)$. This means that when `findNext` returns None, $k \in \mathsf{inr}(n)$ by the traversal invariant and $k \notin \mathsf{outs}(n)$ by the specification of `findNext`. Thus, $k \in \mathsf{ks}(n)$, which by KS-UPD is sufficient to ensure correctness of the decisive operation (the proof is the same as the one we performed for the give-up template).

8.2.2   PROOF OF THE LINK TEMPLATE

We need to make a few more changes to the give-up proof in order to verify the link template. In particular, we need to add a few more types of ghost state to capture some of the special behavior of the link template.

Figure 8.11 contains our definition of the search structure predicate CSS and related predicates that we will use for the link template proof. As with the give-up proof, we use the authori-

$$\mathsf{InFP}(n) := \boxed{\circ\,\{n\}}^{\gamma_f}$$

$$\mathsf{InInr}(k,n) := \exists R.\,\boxed{\circ\,R}^{\gamma_{i(n)}} * k \in R$$

$$\mathsf{N}(n, I_n, C_n) := \mathsf{Node}(n, I_n, C_n) * \boxed{\circ(\mathsf{ks}(I_n, n), C_n)}^{\gamma_k} * \boxed{\tfrac{1}{2}\,I_n}^{\gamma_{h(n)}} * \mathsf{InFP}(n)$$

$$\mathsf{N_S}(n, I_n) := \boxed{\circ I_n}^{\gamma_I} * \boxed{\bullet\mathsf{inr}(I_n, n)}^{\gamma_{i(n)}} * \boxed{\tfrac{1}{2}\,I_n}^{\gamma_{h(n)}} * \mathsf{dom}(I_n) = \{n\}$$

$$\varphi(r, I) := \mathcal{V}(I) \wedge I.in(r) = \mathbb{K} \wedge I.out = \lambda_0$$

$$\mathsf{CSS}(r, C) := \exists I.\,\boxed{\bullet I}^{\gamma_I} * \varphi(r, I) * \boxed{\bullet(\mathbb{K}, C)}^{\gamma_k} * \boxed{\bullet\,\mathsf{dom}(I)}^{\gamma_f}$$
$$* \underset{n \in \mathsf{dom}(I)}{\mathlarger{\mathlarger{\mathlarger{*}}}} \left(\exists b\, I_n\, C_n.\, \mathsf{L}(b, n, \mathsf{N}(n, I_n, C_n)) * \mathsf{N_S}(n, I_n) \right)$$

Figure 8.11: The definition of CSS used by the link template proof.

tative RA of flow interfaces at location γ_I for the flow-based reasoning, the keyset RA from §5.2 to lift local updates to global ones, the authoritative RA of sets of nodes (and the accompanying InFP predicate) to encode the domain of the search structure. But we add two new types of ghost state to this definition.

First, we use fractional RAs at locations $\gamma_{h(n)}$ for each node n to store one half of the node's singleton interface I_n inside N and the other half inside CSS. Since fractional RAs can be updated only when both halves are together, this prohibits other threads from modifying the interface of n when one thread has locked n and removed $\mathsf{N}(n, I_n, C_n)$ from CSS.

Second, we use an authoritative RA of sets of keys, at locations $\gamma_{i(n)}$ for each node n, to encode the inreach of each node. This RA has similar rules as the authoritative RA of sets of nodes at location γ_f (i.e., AUTH-SET-UPD, AUTH-SET-SNAP, and AUTH-SET-UPD). In our proofs, we use these rules so that threads can take snapshots of a node n's inreach and assert that a given key is in it even when they have not locked n. We introduce a new predicate $\mathsf{InInr}(k, n)$ (defined in Figure 8.11) to represent the knowledge a thread has about n's inreach after taking a snapshot as described above. If a thread owns $\mathsf{InInr}(k, n)$, then it knows that k is in the inreach on n. This knowledge is stable under interference of other threads because in the link template, threads only increase the inreach of a node.

Before we describe the link template proof, we present the assumptions it makes about its implementation (summarized in Figure 8.12). Our specifications say that findNext is given a node n satisfying $\mathsf{Node}(n, I_n, C_n)$ and returns None if k is not in the outset of n else $\mathsf{Some}(n')$

$$\{\mathsf{Node}(n, I_n, C_n) * k \in \mathsf{inr}(I_n, n)\}$$
$$\texttt{findNext } n \ k$$
$$\left\{ v. \mathsf{Node}(n, I_n, C_n) * \left(\begin{array}{l} v = \mathsf{None} * k \notin \mathsf{outs}(I_n) \\ \qquad\qquad \vee \ v = \mathsf{Some}(n') * k \in \mathsf{outs}(I_n, n') \end{array} \right) \right\}$$

$$\{\mathsf{Node}(n, I_n, C_n) * k \in \mathsf{ks}(I_n, n)\}$$
$$\texttt{decisiveOp } \omega \ n \ k$$
$$\left\{ v. \mathsf{Node}(n, I_n, C_n') * \left(\begin{array}{l} v = \mathsf{None} * C_n = C_n' \\ \qquad\qquad \vee \ v = \mathsf{Some}(v') * \Psi_\omega(k, C_n, C_n', v') \end{array} \right) \right\}$$

$$\mathsf{Node}(n, I_n, C_n) * \mathsf{Node}(n, I_n', C_n') \twoheadrightarrow \mathsf{False}$$

Figure 8.12: Assumptions the link template proof makes on helper functions and implementation-specific predicates. These are defined and proved by implementations.

such that k is in the outflow to n' (by our definition of edge functions, this means $k \in \mathsf{es}(n, n')$). Similarly, $\texttt{decisiveOp}$ expects a node $\mathsf{Node}(n, I_n, C_n)$ such that k is in the keyset of n. If $\texttt{decisiveOp}$ returns None then it returns the node unchanged. On the other hand, if it returns $\mathsf{Some}(v')$ then the node is now $\mathsf{Node}(n, I_n, C_n')$, and the return value satisfies the search structure specification $\Psi_\omega(k, C_n, C_n', v')$ with respect to the old and new contents of the node n. Finally, we again assume that the heap representation predicate Node cannot be duplicated, hence owning two copies of it implies False.

We now turn to the template proof, shown in Figure 8.13. Recall that our objective is to prove the atomic triple for $\texttt{cssOp}$ from Figure 8.1, using the helper function specifications listed in Figure 8.12 and the lock module specification from Figure 8.2.

The specification of $\texttt{traverse}$ that we prove is almost the same as in the give-up proof, with the addition of $\mathsf{InInr}(k, n)$:

$$\mathsf{InFP}(n) * \mathsf{InInr}(k, n) \twoheadrightarrow$$
$$\left\langle C. \mathsf{CSS}(r, C) \right\rangle \texttt{traverse } r \ n \ k \left\langle v. \mathsf{CSS}(r, C) * \mathsf{N}(v, I_v, C_v) * k \in \mathsf{ks}(I_v, v) \right\rangle$$

This specification says that if n is in the footprint of the search structure, and k is in the inreach of n, then $\texttt{traverse}$ operates on the entire search structure and returns a locked node v such that k is in the keyset of v, while leaving the global contents unmodified.

As before, we start the proof by applying the high-level specification of $\texttt{lockNode}$ from Figure 8.2. This adds $\mathsf{N}(n, I_n, C_n)$ to the thread-local state. Before we can move on to the call to $\texttt{findNext}$, note that $\texttt{findNext}$'s precondition requires $k \in \mathsf{inr}(I_n, n)$ but we have $\mathsf{InInr}(k, n)$. The difference is subtle, but luckily we can convert one to the other using the following lemma that can be proved from the definitions of the involved predicates and AUTH-SET-

```
1  InFP(n) * InInr(k, n) -* ⟨C. CSS(r, C)⟩
2  let rec traverse n k =
3    {InFP(n) * InInr(k, n)}
4    lockNode n;
5    {InFP(n) * InInr(k, n) * N(n, I_n, C_n)}
6    {N(n, I_n, C_n) * k ∈ inr(I_n, n)}
7    match findNext n k with
8    | None ->  {N(n, I_n, C_n) * k ∈ ks(I_n, n)}
9        n
10   | Some n' ->  {N(n, I_n, C_n) * k ∈ inr(I_n, n) * k ∈ outs(I_n, n')}
11       {N(n, I_n, C_n) * InFP(n') * InInr(k, n')}
12       unlockNode n;  {InFP(n') * InInr(k, n')}
13       traverse n' k
14 ⟨v. CSS(r, C) * N(v, I_v, C_v) * k ∈ ks(I_v, v)⟩
15
16 ⟨C. CSS(r, C)⟩
17 let rec cssOp ω r k =
18   {InFP(r) * InInr(k, r)}
19   let n = traverse r k in
20   {N(n, I_n, C_n) * k ∈ ks(I_n, k)}
21   match decisiveOp ω n k with
22   | None ->  {N(n, I_n, C_n)}
23       unlockNode n;  {True}
24       cssOp ω r k
25   | Some res ->
26       {Node(n, I_n, C'_n) * Ψ_ω(k, C_n, C'_n, res) * k ∈ ks(I_n, n) * · · ·}
27       (* Linearization point: open precondition *)
28       ⟨⌈∘(ks(I_n, n), C_n)⌉^{γ_k} * Ψ_ω(k, C_n, C'_n, res) * k ∈ ks(I_n, n) * ⌈•(𝕂, C)⌉^{γ_k} * · · ·⟩
29       ⟨⌈∘(ks(I_n, n), C'_n)⌉^{γ_k} * Ψ_ω(k, C, C', res) * ⌈•(𝕂, C')⌉^{γ_k} * · · ·⟩  (* By KS-UPD *)
30       ⟨N(n, I_n, C'_n) * CSS(r, C') * Ψ_ω(k, C, C', res)⟩
31       unlockNode n;
32       ⟨CSS(r, C') * Ψ_ω(k, C, C', res)⟩  (* Prove postcondition, commit *)
33       {True}
34       res
35 ⟨res. CSS(r, C') * Ψ_ω(k, C, C', res)⟩
```

Figure 8.13: The link template algorithm with a proof outline.

VALID:

$$\text{INInr-INR}$$
$$\mathsf{CSS}(r, C) * \mathsf{N}(n, I_n, C_n) * \mathsf{InInr}(k, n) \twoheadrightarrow k \in \mathsf{inr}(I_n, n)$$

Since we need the CSS to apply INInr-INR, we open the precondition and apply this rule. This is a purely logical step that does not modify the shared state, so we can close the precondition again (i.e., we use AU-ABORT) and get the state shown in line 6. We now have all the resources needed to apply findNext's specification (Figure 8.12).

As before, if findNext succeeds, we directly obtain the postcondition of traverse, so we commit (i.e., this is the linearization point) and complete this branch of the proof. If it fails, then as before we need to prove the precondition of findNext for n' for the recursive call. This uses similar reasoning to the give-up proof, but here we must additionally establish $\text{InInr}(k, n')$. We do this using the following lemma that says that if k is in the inreach of n and is in the outset from n to n', then it must be in the inreach of n':

INREACH-STEP
$$\text{CSS}(r, C) * \text{N}(n, I_n, C_n) * k \in \text{inr}(I_n, n) * k \in \text{outs}(I_n, n') \twoheadrightarrow k \in \text{inr}(k, n')$$

This lemma can be proved using the definition of inreach and some lemmas of the flow framework encoding that we described in §8.2.1. Again, we open the precondition to apply this lemma, closing it again without modifying it. We can then use the high-level specification of unlockNode to return the node n, and obtain the postcondition of traverse by using induction on the recursive call.

The cssOp operation begins with a call to traverse on line 19. To satisfy traverse's precondition, we need to open the precondition and take a snapshot of the global footprint (using AUTH-SET-SNAP and $\varphi(r, I) \Rightarrow r \in \text{dom}(I)$), obtaining $\text{InFP}(r)$. Also, $\varphi(r, I) \Rightarrow k \in \text{inr}(I_r, r)$ so we also take a snapshot of r's inreach at ghost location $\gamma_{i(r)}$ to add $\text{InInr}(k, r)$ to our context. The resulting context is depicted in line 18.

To call traverse we also need $\text{CSS}(r, C)$, so we need to open the precondition again. Since traverse has an atomic triple, it behaves atomically and we can open atomic preconditions (i.e., use AU-ABORT and AU-COMMIT) around calls to it. After traverse returns, we add its postcondition in line 14 to our context (minus $\text{CSS}(r, C)$, which needs to be given back to re-establish cssOp's precondition since we do not commit here). The next step is the call to decisiveOp, for which we already have the precondition in our context.

We then look at the two possible outcomes of decisiveOp. In the case where it returns None, our context is unchanged, so we execute unlockNode using the $\text{N}(n, I_n, C_n)$ in our context. We can use the specification of cssOp on the recursive call on line 24 to complete this branch of the proof.

On the other hand, if decisiveOp succeeds, we get back a modified node $\text{Node}(n, I_n, C_n')$ with new contents C_n' that satisfies the search structure specification $\Psi_\omega(k, C_n, C_n', res)$ locally (line 26). We now need to show that this modification results in cssOp's postcondition; this is essentially the *linearization point* of this algorithm.

To do this, we again open the atomic precondition $\text{CSS}(r, C)$. We now have the context in line 28 (we have also expanded $\text{N}(n, I_n, C_n')$), and now we can apply our ghost update KS-UPD to update the global contents and get the context in line 29. In particular, we have $\Psi_\omega(k, C, C', res)$

and $\mathsf{CSS}(r, C')$, which allows us to "commit" and establish the postcondition. We finally apply the specification of `unlockNode` using the remaining $\mathsf{N}(n, I_n, C'_n)$, and complete the proof.

8.2.3   MAINTENANCE OPERATIONS.

As with the give-up template, maintenance operations for the link technique need to be proved separately. The specification for a maintenance operation `maintOp` is the same:

$$\big\langle C.\ \mathsf{CSS}(r, C) \big\rangle \mathtt{maintOp}\ r\ \big\langle \mathsf{CSS}(r, C) \big\rangle$$

Again, this triple gives the maintenance operation the permission to modify the search structure $\mathsf{CSS}(r, C)$ as long as it appears to instantaneously modify the structure to another valid search structure $\mathsf{CSS}(r, C)$ with the same contents. While we once again omit the full proofs here, we describe them at a high level below and show the new lemmas needed.

For the B-link tree, the maintenance operations are the half-split, full-split, and free-at-empty. The interesting part of their proofs is in showing that they do not decrease the inreach of any node, which for the half-split and free-at-empty is easy to do. The half-split also requires us to reason about the creation of a new node, which we do using the domain extension notion (Definition 8.1) and frame-preserving update from §8.1.2.

For the full-split, we have discussed in §8.2.1 how we need a frame-preserving update that allows increasing the inflow of nodes that allows us to show that the full-split preserves the inreach of all modified nodes. This is formally justified by a notion of interface extension that allows increasing the inflow of the modified region. Note that this definition is only for flow domains that are positive monoids (see Chapter 2), which is true of multisets of keys, the flow domain that we use.

Definition 8.2   Given a positive flow domain, a flow interface I' is an *inflow extension* of interface I, written $I \preccurlyeq I'$, if and only if

1. $\mathsf{dom}(I) = \mathsf{dom}(I')$,

2. $\forall n \in I.\ I.in(n) \leq I'.in(n)$, and

3. $I.out = I'.out$.

This definition allows I' to differ from I by having a larger inflow, as long as the domains and the outflow of the modified region are exactly the same. Similarly, we can show that replacing an interface with an inflow extension is a frame-preserving update in the flow interface RA:

$$\frac{\text{FLOWINT-INF-UPD}}{I_1 \preccurlyeq I'_1 \qquad \text{the flow domain is positive}}{(\bullet I_1 \oplus I_2, \circ I_1) \rightsquigarrow \{(\bullet I'_1 \oplus I_2, \circ I'_1) \mid (I_1 \oplus I_2) \preccurlyeq (I'_1 \oplus I_2)\}}$$

We use this lemma in the full-split to update the flow interface ghost state in such a way as to preserve the inreach of all modified nodes.

```
1  let create () =                          12  let rec searchStrOp ω r k =
2    let r = allocRoot () in                13    lockNode r;
3    r                                      14    match findNext r k with
4                                           15    | None -> searchStrOp ω r k
5  let rec traverse p n k =                 16    | Some n ->
6    match findNext n k with                17      let (p, n) = traverse r n k in
7    | None -> (p, n)                       18      let res = decisiveOp ω p n k in
8    | Some n' ->                           19      unlockNode p;
9      unlockNode p;                        20      unlockNode n;
10     lockNode n;                          21      res
11     traverse n n' k
```

Figure 8.14: The lock-coupling template algorithm. Like all our templates, the main method is cssOp and traverse is an auxiliary method used for recursive traversal.

8.3   THE LOCK-COUPLING TEMPLATE

The lock-coupling template (Figure 8.14) uses the hand-over-hand locking scheme to ensure that no thread is negatively interfered with while it is on its traversal. Unlike the other two templates in this chapter, every thread always holds at least one lock while traversing from one node to the next. This means that no other thread can overtake this thread, or perform any modification that would invalidate this thread's search.

This template can be instantiated to an implementation that uses a sorted singly linked list, where each node contains a single key. Insert operations create a new node and link the node into the appropriate position of the list, while delete operations unlink the node containing the operation key from the list by swinging the pointer from the predecessor node to the successor node (as shown in Figure 7.1).

The lock-coupling technique is a simpler and less efficient concurrency technique. It can be proved, on paper at least, using the standard conflict-preserving serializability technique [Bernstein et al., 1987]. Thus, we omit the details of its proof using our technique here. The full proof is available online in our public repository of machine-checked proofs (see Chapter 13), along with the other proofs in our monograph.

8.4   VERIFYING IMPLEMENTATIONS

To obtain a verified implementation of one of the templates, one needs to specify the concrete representation of a node by defining the Node predicate and provide code for the helper functions that satisfies the specifications assumed by the template in question. For example, to verify an implementation of the give-up template, one needs to provide implementations for Node, inRange, findNext, and decisiveOp that satisfy the assumptions listed in Figure 8.5.

We would like to re-emphasize that these specifications use sequential Hoare triples and assert ownership of only locked nodes. Thus, if their implementations are sequential code, we can verify them using an off-the-shelf separation logic tool that can verify sequential heap-manipulating code. Such tools typically provide better automation, and can help speed up the verification process.

Chapter 13 discusses the tools we used and the implementations we verified for each of the three template algorithms seen in this chapter.

CHAPTER 9

Verifying Multicopy Structures

In Chapter 8, we demonstrated how to simplify the verification of concurrent search structures by abstracting concurrency algorithms underlying diverse implementations such as B-trees and hash tables into templates that can be verified once and for all. The template algorithms we have considered so far handle only search structures that perform all operations on keys *in-place*. That is, an operation on key k searches for the unique node containing k in the structure and then performs any necessary modifications on that node. Since every key occurs at most once in the data structure at any given moment, we refer to these structures as *single-copy structures*.

Single-copy structures achieve high performance for reads. However, some applications, such as event logging, require high write performance, possibly at the cost of decreased read speed and increased memory overhead. This demand is met by data structures that store upserts (inserts, deletes or updates) to a key k *out-of-place* at a new node instead of overwriting a previous copy of k that was already present in some other node. Performing out-of-place upserts can be done in constant time (e.g., always at the head of a list). A consequence of this design is that the same key k can now be present multiple times simultaneously in the data structure. Hence, we refer to these structures as *multicopy structures*.

Examples of multicopy structures include the differential file structure [Severance and Lohman, 1976], the log-structured merge (LSM) tree [O'Neil et al., 1996], and the Bw-tree [Levandoski et al., 2013]. These concurrent data structures are widely used in practice, including in state-of-the-art database systems such as Apache Cassandra [Apache Software Foundation, 2020a] and Google LevelDB [Google, 2020]. The differential file structure and LSM tree, in particular, can be tuned by implementing workload- and hardware-specific data structures at the node level. Despite the differences between these implementations, they generally follow the same high-level template algorithms for the core search structure operations.

The nodes of multicopy data structures contain their own data structures, typically a simple array at the root to allow upserts to perform fast appends and a classical single-copy search structure (e.g., a hash structure or arrays with bloom filters) for non-root nodes. The non-root nodes are typically read-only, so concurrency at the node level is not an issue. In this monograph, we consider the multicopy data structure as a graph of nodes and study template algorithms on that graph.

Figure 9.1 provides an overview of the development that we will present in the next chapters. In the remainder of this chapter, we will first describe the basic intuition behind the correctness proof of any multicopy structure. Next, we will introduce the differential file structure [Sev-

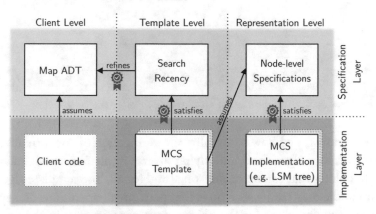

Figure 9.1: The structure of our verification effort. MCS stands for Multicopy Structure and LSM stands for Log-structured Merge.

erance and Lohman, 1976] and LSM tree [O'Neil et al., 1996] in more detail. In Chapter 10, we then derive an abstract notion of multicopy structures similar to the abstract single-copy structures in the edgeset framework. The new framework increases proof reusability by introducing an intermediate abstraction level at which we can reason about concurrent template algorithms for multicopy structures without committing to a concrete representation of the data structure in memory.

From the perspective of client code, a linearizable multicopy structure behaves like a Map Abstract Data Type (ADT). To simplify the linearizability proof for this client-level specification, we introduce an intermediate specification at the template level, called *search recency*. This specification tracks sufficient information about the history of concurrent executions so that one only needs to reason about fixed local linearization points when verifying specific multicopy structure algorithms. We then show in Chapter 11 that any concurrent multicopy structure that obeys search recency is linearizable with respect to the client-level specification.

Finally, in Chapter 12 we discuss template algorithms for concurrent multicopy structures and prove that they satisfy the desired template-level specification. As for our single-copy structure templates, the multicopy structure templates and proofs abstract from the concrete representation of the data structure in memory, supporting implementations based on different heap representations such as lists, arrays, and B-link trees. This enables proof reuse across diverse implementations.

9.1   A LIBRARY ANALOGY TO MULTICOPY STRUCTURES

To train your intuition, consider a library of books in which new editions of the same book arrive over time. Thus, the first edition of book k can enter and later the second edition, then the third, and so on. A patron of this library who enters the library at time t and is looking for book k

should find an edition that is either current at time t or one that arrives in the library after t. We call this normative property *search recency*.

Now suppose the library is organized as a sequence of rooms. All new books are put in the first room (near the entrance). When a new edition v of a book arrives in the first room, any previous editions of that book in that room are thrown out. When the first room becomes full, some of the books in that room are moved to the second room. If a previous edition of some book is already in the second room, that previous edition is thrown out. When the second room becomes full, then some of its books are moved to the third room using the same throwing out rule, and so on. This procedure maintains the time-ordering invariant that the editions of the same book are ordered from most recent (at or nearer to the first room) to least recent (farther away from the first room) in the sequence of rooms.

A patron's search for k starting at time t begins in the first room. If the search finds any edition of k in that room, the patron takes a photocopy of that edition. If not, the search proceeds to the second room, and so on.

Now suppose that the latest edition at time t is edition v and there is a previous edition v'. Because of the time-ordering invariant and the fact that the search begins at the first room, the search will encounter v before it encounters v'. The search may also encounter an even newer edition of k, but will never encounter an older one before returning. That establishes the search recency property.

Any concurrent execution of inserts and searches is equivalent to a serial execution in which (i) each insert is placed in its relative order of entering the root node with respect to other inserts and (iia) a search s is placed after the insert whose edition s copies if that insert occurred after s began or (iib) a search s is placed at the point when s began, if the edition that s copies was inserted before s began (or if s returns no edition at all).

Because the searches satisfy the search recency property, the concurrent execution is linearizable, which is our ultimate correctness goal.

9.2   DIFFERENTIAL FILE STRUCTURES

The differential file structure is one of the earliest examples of a key-value store that allows out-of-place updates. It closely corresponds to the library analogy described above where the library consists of exactly two rooms. The structure stores the data in a *main file* on disk (the second room of the library). However, when key-value pairs are upserted, the changes are first recorded in a *differential file* in memory (the first room) instead of modifying the main file directly. In the following, we will refer to the differential file as the *root node* and the main file as the *disk node* of the data structure. Figure 9.2 shows a potential state of a differential file structure with root node r and disk node n.

When searching for the value associated with a given operation key k, the search thread consults the root node first. If k is not found, then the thread proceeds to the disk node. For

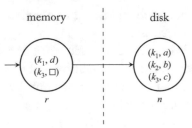

Figure 9.2: A differential file structure.

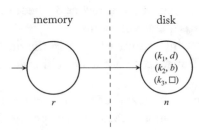

Figure 9.3: A differential file structure obtained after reorganizing data in Figure 9.2.

instance, a search for key k_2 on the differential file structure in Figure 9.2 will return b, while a search for k_3 will return $\square$ indicating that k_3 has been deleted.

Periodically (e.g., after a certain time period or when the root node becomes full), a maintenance operation reorganizes the structure by moving the contents of the root node to the disk node, leaving the root node empty so it can begin collecting changes once again. In case of a conflict when moving the data, i.e., if a copy of key k exists in both nodes, then the copy in the root node is kept to preserve the time-ordering invariant. Figure 9.3 shows the structure obtained after reorganizing the structure shown in Figure 9.2. Here, the copy $(k_3, \square)$ in the root node is kept over (k_3, c) in the disk node.

Our development in Iris includes a two-node multicopy structure template that captures the operations on the differential file structure. However, we do not discuss this template in detail in this monograph and instead proceed directly with the more general Log-Structured Merge tree.

9.3   LOG-STRUCTURED MERGE TREES

The LSM tree targets applications that require high write performance such as file systems for storing transactional log data. The LSM tree can be seen as a generalization of the differential file structure, in the sense that it replaces the single disk node n with a list of disk nodes $n_1, n_2, \ldots, n_l$. Figure 9.4 shows an example. As in the case of differential file structures, the implementation of the node-level structures can depend on the storage medium. Some modern

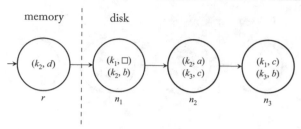

Figure 9.4: High-level structure of an LSM tree.

implementations use skip lists for memory nodes and sorted string tables of fixed size for the nodes stored on disk[1].

The LSM tree operations essentially behave as outlined in the library analogy. The upsert operation takes place at the root node r. A search for a key k traverses the list starting from the root node and retrieves the value associated with the first copy of k that is encountered. If the retrieved value is $\square$ or if no entry for k has been found after traversing the entire list, then the search determines that k is not present in the data structure. Otherwise, it returns the retrieved value. For instance, a search for key k_1 on the LSM tree depicted in Figure 9.4 would determine that this key is not present since the retrieved value is $\square$ from node n_1. Similarly, k_4 is not present since there is no entry for this key. On the other hand, a search for k_2 would return d and a search for k_3 would return c.

To prevent the root node from growing too large, the LSM tree performs *flushing*. As the name suggests, the flushing operation flushes the data from the root node to the disk by moving its contents to the first disk node. Figure 9.5 shows the LSM tree obtained from Figure 9.4 after flushing the contents of r to the disk node n_1.

Similar to flushing, a *compaction* operation moves data from full nodes on disk to their successor. In case there is no successor, a new node is created at the end of the structure. During the merge, if a key is present in both nodes, then the most recent (closer-to-the-root) copy is kept, while older copies are discarded. Figure 9.6 shows the LSM tree obtained from Figure 9.4 after compacting nodes n_1 and n_2. Here, the copy of k_2 in n_2 has been discarded. In practice, the length of the data structure is bounded by letting the size of newly created nodes grow exponentially.

As is the case for differential file structures, the net effect of all these operations is that the data structure satisfies the time-ordering invariant and searches achieve search recency.

The LSM DAG template discussed in Chapter 12 generalizes the operations on the LSM tree to arbitrary directed acyclic graphs (DAGs). In the proof of the template, we show how to use the flow framework to capture the time-ordering invariant on DAGs of unbounded size

---

[1]RocksDB [Facebook, 2020], LevelDB [Google, 2020], and Apache HBase [Apache Software Foundation, 2020b] all use variants of concurrent skip list for in-memory data structure and SSTables for disk storage.

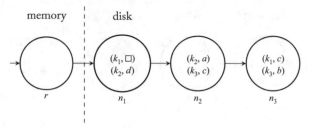

Figure 9.5: LSM tree obtained from Figure 9.4 after flushing node r to disk.

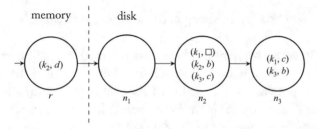

Figure 9.6: LSM tree obtained from Figure 9.4 after compacting nodes n_1 and n_2.

while preserving the strategy of decomposing the proof into concurrency-related aspects and the heap-related aspects as we did for the single-copy structure templates discussed in Chapter 8.

CHAPTER 10

The Edgeset Framework for Multicopy Structures

In this chapter, we present the edgeset framework for multicopy structures and formally define the specifications that each operation on a multicopy structure must satisfy.

10.1 MULTICOPY STRUCTURES FORMALLY

Analogously to the verification methodology described in Chapter 8, we abstract away from the data organization within the nodes, and treat the data structure as consisting of nodes in a directed acyclic graph. Furthermore, we will abstract from the concrete data values associated with the keys in the data structure to keep the presentation simple.

Since copies of a single key k can be present in different nodes simultaneously, we need a mechanism to differentiate between these copies. To that end, we represent the entries of the contents of the data structure as pairs (k, t) where t is a timestamp uniquely identifying the point in time when this copy of k was upserted. We use a single global clock for this purpose. For example in Figure 10.1, $(k_3, 4)$ was upserted after $(k_2, 3)$, which was upserted after $(k_3, 1)$. Note that the timestamp associated with a key is auxiliary, or ghost, data that we use in our proofs to track the temporal ordering of the copies present in the structure at any point. Implementations do not need to explicitly store this timestamp information.

Formally, let $\mathbb{K}$ be the set of all keys. A multicopy structure is a directed acyclic graph $G = (N, E)$ with nodes N and edges $E \subseteq N \times N$. We assume that there is a dedicated *root node* $r \in N$ which uniquely identifies the structure. Each node n of the graph is labeled by its contents $C_n: \mathbb{K} \to \mathbb{N}_\perp$, which is a map from keys to timestamps ($\mathbb{N}_\perp := \mathbb{N} \uplus \{\perp\}$). For a node n and its contents C_n, we say (k, t) is in the contents of n if $C_n(k) = t$. We denote the absence of key k in n by $C_n(k) = \perp$ and let $\mathsf{dom}(C_n) := \{k \mid C_n(k) \neq \perp\}$. For each edge $(n, n') \in E$ in the graph, the edgeset $\mathsf{es}(n, n')$ is the set of keys k for which an operation arriving at a node n would traverse (n, n') if $k \notin \mathsf{dom}(C_n)$. We require that the edgesets of all outgoing edges of a node n be pairwise disjoint. Figure 10.1 shows a potential abstract multicopy structure graph consistent with the LSM tree depicted in Figure 9.4. Here, all edges have edgeset $\mathbb{K}$.

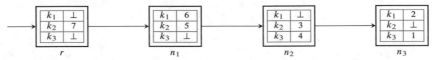

Figure 10.1: Abstract multicopy data structure graph for the LSM tree in Figure 9.4.

10.2 CLIENT-LEVEL SPECIFICATION

Our goal is to prove the linearizability of concurrent multicopy structure templates with respect to their desired sequential client-level specification. As discussed earlier, the sequential specification is that of a map ADT, i.e., the logical contents of the data structure is a partial mathematical map from keys to values.

Here, we view a multicopy structure abstractly as a map from keys to timestamps. When proving linearizability, we will place searches in an equivalent sequential history based on the timestamp that they return. (The timestamp plays the role of the book edition in the library analogy from the last chapter.) This simplifies the correctness argument. One can easily recover a specification in terms of key/value maps by tracking the upserted values along with the timestamps of the upsert. (We omit values from our discussion because the existence and form of values is irrelevant to the core of the proof.)

To formalize the above idea, we extend the ordering on natural numbers to $\mathbb{N}_\perp$ by defining $\perp < n$ for all $n \in \mathbb{N}$. The *logical (timestamp) contents* of a multicopy structure is the map $M: \mathbb{K} \to \mathbb{N}_\perp$ that associates every key with the timestamp of its latest copy in the structure:

$$M(k) := \max \{t \mid \exists n \in N.\ C_n(k) = t\}$$

We call $M(k)$ the *logical timestamp* of key k. Note that $M(k) = \perp$ indicates that no copy of k exists in the structure, i.e., k has not been upserted so far.

Now suppose that $\overline{\mathrm{MCS}}(r, t, M)$ is a representation predicate that provides the client view of a multicopy structure with root r, abstracting its shared state by the logical contents M as well as the current value t of the global clock. We then require that the search and upsert methods respect the following client-level atomic specifications:

$$\big\langle t\, M.\, \overline{\mathrm{MCS}}(r, t, M) \big\rangle\ \text{upsert}\ r\ k\ \big\langle \overline{\mathrm{MCS}}(r, t+1, M[k \rightarrowtail t]) \big\rangle \tag{10.1}$$

$$\big\langle t\, M.\, \overline{\mathrm{MCS}}(r, t, M) \big\rangle\ \text{search}\ r\ k\ \big\langle t'.\ \overline{\mathrm{MCS}}(r, t, M) * M(k) = t' \big\rangle \tag{10.2}$$

The specification of upsert updates the logical timestamp of k to the current global clock value and increments the global clock. Thus, upsert performs the "insert" of the library analogy. The search specification states that search returns the logical timestamp t' of its operation key.

10.3   TEMPLATE-LEVEL SPECIFICATION: SEARCH RECENCY

The verification of multicopy structures requires reasoning about the dynamic non-local linearization points of search, which are determined by the concurrently executing upserts. We want to avoid having to do this reasoning each time we verify a new template for a multicopy structure implementation. Our strategy is to provide an alternative template-level specification that uses a more detailed abstraction of the computation history rather than just the logical contents. This alternative specification will then have fixed local linearization points, simplifying the verification.

We say that search satisfies *search recency* if each concurrent invocation search r k either returns the logical timestamp associated with k at the point when the search started, or any other copy of k that was upserted between the search's start time and the search's end time.

We will show that if searches satisfy search recency and upserts take effect in a single atomic step that changes the logical contents M according to (10.1), then the multicopy structure is linearizable.

To see that search recency is also a necessary condition for linearizability, consider a concurrent execution where a thread T_1 completes an upsert of key k_2 at timestamp 7, resulting in the multicopy structure depicted in Figure 10.1. Suppose that after T_1 returns another thread T_2 starts a search for k_2. If we relaxed the search recency condition to allow older copies of keys, T_1 would, e.g., be allowed to return the copy of k_2 associated with timestamp 5. However, the logical timestamp of k_2 throughout the execution of search is never 5. In order to obtain a linearization of this execution that is consistent with the client-level specification, we would have to reorder the two calls so that T_1's call to search completes before T_2's call to upsert. This would violate linearizability since the invocation point of T_2's search is after the return point of T_1's upsert.

A formal specification of search that captures the search recency property needs to relate the return value of search to the copies of the operation key k that were concurrently upserted while the search executed. Our abstraction of a multicopy structure in terms of its logical contents M is not well suited for this purpose because it loses too much information about the data structure's state: the logical contents tracks only the latest copies of keys but not older copies that were previously upserted and that an ongoing search may still access.

We therefore first move to a slightly lower level abstraction. To this end, we define the *upsert history* $H \subseteq \mathbb{K} \times \mathbb{N}_\perp$ of a multicopy data structure as the set of all copies (k, t) that have been upserted thus far. In particular, this means that any multicopy structure will have to maintain the invariant $H \supseteq \bigcup_{n \in N} C_n$. We further define $\bar{H} := \lambda k. \max \{t \mid (k, t) \in H\}$. Since the data structure must always contain the latest copy of a key that was upserted, $\bar{H}$ coincides with the logical contents M.

Assume that, similar to $\overline{\mathrm{MCS}}(r, t, M)$, we are given a template-level representation predicate $\mathrm{MCS}(r, t, H)$ that abstracts the state of a multicopy structure by its upsert history H and the current value t of the global clock. The desired template-level specification of `upsert` in terms of the new abstraction is simply:

$$\langle t\, H.\, \mathsf{MCS}(r, t, H)\rangle\, \texttt{upsert}\ r\ k\ \langle \mathsf{MCS}(r, t+1, H \cup (k, t))\rangle \tag{10.3}$$

The postcondition of `search` needs to express two properties. First, we must necessarily have $(k, t') \in H$, where t' is the returned timestamp and H is the value of the upsert history at the linearization point. In addition, if H_0 is the value of the upsert history at the start of the search, we need to capture that t' is either the logical timestamp t_0 of k at that point (i.e., $t_0 := \bar{H}_0(k)$) or t' is the timestamp of an upsert for k that happened after the search started, i.e., $t_0 < t'$. This is equivalent to demanding that for all t_0' such that $(k, t_0') \in H_0$, the returned timestamp t' satisfies $t_0' \leq t'$. We define the auxiliary abstract predicate $\mathsf{SR}(k, t)$ to mean that $(k, t) \in H$ for the value H of the upsert history at the time point when the predicate is evaluated[1]. Using this predicate, the template-level specification of `search` is then expressed as follows:

$$\begin{aligned}&\forall t_0'.\, \mathsf{SR}(k, t_0') \twoheadrightarrow \\ &\quad \langle t\, H.\, \mathsf{MCS}(r, t, H)\rangle\, \texttt{search}\ r\ k\ \langle t'.\, \mathsf{MCS}(r, t, H) * t_0' \leq t' * (k, t') \in H\rangle\end{aligned} \tag{10.4}$$

Here, we use the magic wand connective $\twoheadrightarrow$ to express that the auxiliary local precondition $\mathsf{SR}(k, t_0')$ must be satisfied at the time point when `search` is invoked.

In the next chapter, we deal with the complexity of non-local dynamic linearization points of searches once and for all by proving that any multicopy structure that satisfies the template-level specification also satisfies the desired client-level specification. To prove the correctness of a given concurrent multicopy structure, it then suffices to show that `upsert` satisfies its corresponding template-level specification (10.3) and `search` satisfies (10.4). When proving the validity of the template-level atomic triple of `search` for a particular implementation (or template), one can now always *commit* the atomic triple (i.e., declare a linearization point) at the point when the return value of the search is determined, i.e., when $(k, t') \in H$ is established.

Maintenance Operations.   For a maintenance operation `maintOp` (e.g., `flush` or `compact` for the LSM tree) the sequential and atomic specifications simply demand that the operation does not change the abstract state of the data structure:

$$\langle t\, M.\, \overline{\mathsf{MCS}}(r, t, M)\rangle\, \texttt{maintOp}\ r\ \langle \overline{\mathsf{MCS}}(r, t, M)\rangle \tag{10.5}$$

$$\langle t\, H.\, \mathsf{MCS}(r, t, H)\rangle\, \texttt{maintOp}\ r\ \langle \mathsf{MCS}(r, t, H)\rangle \tag{10.6}$$

Note that we do not allow maintenance operations to change the clock value t since the global clock corresponds to the number of upsert operations completed so far. Since maintenance operations do not affect the abstract state of the data structure, the development in the remainder of this chapter can be adapted to accommodate maintenance operations.

---

[1]In §11.4 we will express $\mathsf{SR}(k, t)$ using appropriate Iris ghost state that keeps track of the upsert history.

CHAPTER 11

Reasoning about Non-Static and Non-Local Linearization Points

In this chapter, we prove that any concurrent execution of upsert and search operations that satisfy the template-level specifications (10.3) and (10.4) can be linearized to an equivalent sequential execution that satisfies the client-level specifications. Intuitively, this can be done by letting the upserts in the equivalent sequential execution occur in the same order as their atomic commit points in the concurrent execution, and by letting each search r k occur at the earliest time after the timestamp referenced by the returned copy t' of k. That is, if $t' = t_0$ (recall that t_0 is the logical timestamp of k at the start of the search on k), then the search occurs right after it was invoked in the concurrent execution. Otherwise, we must have $t' > t_o$ and the search occurs after the upsert at time t'. The fact that such an upsert must exist follows from the template-level specifications.

The intuitive proof argument above relies on explicit reasoning about execution histories. Instead, we aim for a thread-modular proof that reasons about individual searches and upserts in isolation, so that we can mechanize the proof in a Hoare logic like Iris. The proof we present below takes inspiration from that of the RDCSS data structure presented in Jung et al. [2020].

11.1 CHALLENGES AND PROOF OUTLINE

There are two obstacles that we need to overcome in order to obtain a thread-modular proof. The first obstacle is that the linearization points of search are non-static. That is, a thread executing search does not know beforehand whether it will find a timestamp that is the logical value at the start of the search, or one that has been concurrently upserted later. Hence, the thread does not know whether to linearize immediately when the search starts, or to wait for the completion of a later upsert. Here, Iris' *prophecies* come to our rescue [Jung et al., 2020]. Prophecies, which were first introduced in Abadi and Lamport [1988], allow a thread to predict what will happen in the future. In particular, one can use prophecies to predict future events in order to reason about non-fixed linearization points [Vafeiadis, 2008, Zhang et al., 2012]. In our case, a thread executing search can use a prophecy to predict, at the beginning of the search, the timestamp

t' that it will eventually return. In a thread-modular correctness proof one can then decide on how to linearize the operation based on the predicted value.

The second obstacle is that linearization points of search are also *non-local*. Specifically, in the case $t_0 < t'$, i.e., where a search operation does not linearize immediately, the linearization point occurs when an instruction of a concurrent upsert is executed. One can view this as a form of *helping* [Liang and Feng, 2013]: when an upsert operation commits and adds (k, t') to the upsert history H, it also commits all the (unboundedly many) concurrently executing search operations for k that will return timestamp t'. We encode this *helping protocol* in the predicate $\overline{MCS}(r, t, M)$ that captures the shared (ghost) state of the data structure, by taking advantage of Iris' support for higher-order ghost state.

At a high level, the proof then works as follows. For proving the client-level atomic triple of search from the template-level specification, we augment search with auxiliary ghost code that creates and resolves the relevant prophecies. At the start of a search for key k, the current thread creates a fresh prophecy that predicts the returned timestamp t'. If t' is equal to the current logical value t_0 of k, then the thread commits the client-level atomic triple right away. If instead $t' > t_0$, then the thread *registers* its own client-level atomic triple in the shared predicate $\overline{MCS}(r, t, M)$. The registered atomic triple serves as an obligation to commit the atomic triple. This obligation will be discharged by the upsert operation adding (k, t') to H. The search thread then uses the template-level specification of search to conclude that it can collect the committed triple from the shared predicate after search has returned.

Relating the high-level and low-level specification of upsert is straightforward. However, the proof of upsert also needs to do its part of the helping protocol by scanning over all the searches that are currently registered in the shared predicate $\overline{MCS}(r, t, M)$ and committing those that return the copy of k added by the upsert.

In the remainder of this chapter we explain this *helping proof* in more detail.

11.2   ADDING PROPHECIES

```
1 let search r k =
2   let tid = NewProph in
3   let p = NewProph in
4   let t' = search r k in
5   Resolve p to t';
6   t'
```

ONE-SHOT-PROPHECY-CREATION
$$\{True\}\ \mathtt{NewProph}\ \{p.\ \exists vp.\ \mathsf{Proph}(p, vp)\}$$

ONE-SHOT-PROPHECY-RESOLUTION
$$\{\mathsf{Proph}(p, vp)\}\ \mathtt{Resolve}\ p\ \mathtt{to}\ v\ \{vp = v\}$$

Figure 11.1: Wrapper augmenting search with prophecy-related ghost code. The right side shows the specifications of the ghost code.

As explained earlier, we need to augment search with ghost code that creates and resolves the relevant prophecies. We do this by defining the wrapper function $\overline{search}$ given in Figure 11.1.

The right side shows the specifications of the two functions related to manipulating (one-shot) prophecies in Iris. The function NewProph returns a fresh prophecy p that predicts the value vp. This fact is captured by the resource $\mathsf{Proph}(p, vp)$ in the postcondition of the Hoare triple specifying NewProph. The wrapper function $\overline{\text{search}}$ uses NewProph to create two prophecies that it binds to tid and p. The prophecy p predicts the value t' that will eventually be returned by search. The value predicted by the second prophecy tid will be irrelevant. We will use tid later as a unique identifier of the thread performing the search when we encode the helping protocol, taking advantage of the fact that each prophecy returned by NewProph is fresh.

The resource $\mathsf{Proph}(p, vp)$ can be owned by a thread as well as transferred between threads via shared resources such as the representation predicate $\mathsf{MCS}(r, t, H)$ (as is usual in concurrent separation logics). Additionally, $\mathsf{Proph}(p, vp)$ is an exclusive resource, meaning it cannot be duplicated:

PROPH-EXCLUSIVE
$$\mathsf{Proph}(p, \_) \twoheadrightarrow \mathsf{Proph}(p, \_) \twoheadrightarrow \mathsf{False}$$

This ensures that the prophecy can be resolved only once, which is done using Resolve p to v. This operation consumes the resource $\mathsf{Proph}(p, vp)$ and yields the proposition $vp = v$. It is used on line 5 of $\overline{\text{search}}$ to express that the value predicted by p is indeed the value t' returned by $\overline{\text{search}}$.

The erasure theorem for prophecy-related ghost code [Jung et al., 2020, Theorem 3.2] implies that it is sound to treat calls to search as calls to $\overline{\text{search}}$ when reasoning about a client of the multicopy structure.

11.3   DECOUPLING THE HELPING AND TEMPLATE PROOFS

In our proofs, we parameterize the representation predicate $\overline{\text{MCS}}$ by two abstract predicates $\mathsf{Inv}_{tpl}(r, t, H)$ and $\mathsf{Prot}(H)$. We do this to achieve a complete decoupling of the template-specific proofs from the helping proof, which relates the client-level and template-level specifications, and vice versa. The predicate $\mathsf{Inv}_{tpl}(r, t, H)$ abstracts from the resources needed for proving that a particular multicopy structure template satisfies the template-level specifications. In particular, this predicate will store the authoritative version of the global clock t. The predicate $\mathsf{Prot}(H)$ abstracts from the resources used to track the state of the helping protocol.

The fact that Prot depends on H creates an unfortunate entanglement between the proofs performed at the two abstraction levels: at the linearization point of upsert, the upsert history must be updated from H to $H \cup \{(k, t)\}$. This step happens in the proof of each particular template algorithm for upsert. Therefore, these proofs will also have to carry out the ghost update that replaces $\mathsf{Prot}(H)$ by $\mathsf{Prot}(H \cup \{(k, t)\})$.

Fortunately, the template-specific proof does not need to know how Prot is defined. It only needs to know that for any values of H, k, and t, $\mathsf{Prot}(H)$ can be updated to $\mathsf{Prot}(H \cup \{(k, t)\})$.

We capture this assumption on Prot formally using a predicate Updatable(Prot) that we define using Iris' linear view shift modality:

$$\text{Updatable}(\text{Prot}) := (\forall\ H\ k\ t.\ \text{Prot}(H) \Rrightarrow \text{Prot}(H \cup \{(k,t)\}))$$

In order to verify a particular implementation of `search` and `upsert` with respect to a particular template-specific invariant $\text{Inv}_{tpl}(r,t,H)$, one then needs to prove validity of the following Iris propositions:

$$\forall\ \text{Prot}\ r\ k.\ \boxed{\text{Inv}(\text{Inv}_{tpl}, \text{Prot})(r)} \twoheadrightarrow \text{Updatable}(\text{Prot}) \twoheadrightarrow$$
$$\langle t\ H.\ \text{MCS}(r,t,H) \rangle\ \texttt{upsert}\ r\ k\ \langle \text{MCS}(r,t+1,H \cup (k,t)) \rangle \tag{11.1}$$

$$\forall\ \text{Prot}\ r\ k\ t_0.\ \boxed{\text{Inv}(\text{Inv}_{tpl}, \text{Prot})(r)} \twoheadrightarrow \text{SR}(k,t_0) \twoheadrightarrow$$
$$\langle t\ H.\ \text{MCS}(r,t,H) \rangle\ \texttt{search}\ r\ k\ \langle t'.\ \text{MCS}(r,t,H) * t_0 \leq t' * (k,t') \in H \rangle \tag{11.2}$$

These propositions abstract over any helping protocol predicate Prot that is compatible with the ghost update of H performed by `upsert`. Note that an atomic triple guarded by an invariant can be interpreted as satisfying the atomic triple under the assumption that the shared state satisfies the invariant.

We show in the next section that there exists a specific helping protocol invariant Prot_{help} such that the client-level atomic triples hold for any implementation of `upsert` that satisfies proposition 11.1 and any implementations of `search` that satisfies proposition 11.2. Formally, if we abbreviate proposition 11.1 by $\text{UpsertSpec}(\texttt{upsert}, \text{Inv}_{tpl})$ and proposition 11.2 by $\text{SearchSpec}(\texttt{search}, \text{Inv}_{tpl})$. Then we show validity of the following two propositions:

$$\forall\ \texttt{upsert}\ \text{Inv}_{tpl}\ r\ k.\ \text{UpsertSpec}(\texttt{upsert}, \text{Inv}_{tpl}) \twoheadrightarrow$$
$$\langle t\ M.\ \overline{\text{MCS}}(\text{Inv}_{tpl}, \text{Prot}_{help})(r,t,M) \rangle$$
$$\texttt{upsert}\ r\ k \tag{11.3}$$
$$\langle \overline{\text{MCS}}(\text{Inv}_{tpl}, \text{Prot}_{help})(r,t+1,M[k \rightarrowtail t]) \rangle$$

$$\forall\ \texttt{search}\ \text{Inv}_{tpl}\ r\ k.\ \text{SearchSpec}(\texttt{search}, \text{Inv}_{tpl}) \twoheadrightarrow$$
$$\langle t\ M.\ \overline{\text{MCS}}(\text{Inv}_{tpl}, \text{Prot}_{help})(r,t,M) \rangle$$
$$\texttt{search}\ r\ k \tag{11.4}$$
$$\langle t'.\ \overline{\text{MCS}}(\text{Inv}_{tpl}, \text{Prot}_{help})(r,t,M) * M(k) = t' \rangle$$

These propositions abstract over the template-specific parameters `upsert`, `search`, and Inv_{tpl}, but fix the helping protocol invariant Prot to be Prot_{help}. In order to obtain the overall correctness proof of a specific template (`upsert`, `search`, Inv_{tpl}), the proofs of these generic propositions then need to be instantiated only with the proofs of $\text{UpsertSpec}(\texttt{upsert}, \text{Inv}_{tpl})$ and $\text{SearchSpec}(\texttt{search}, \text{Inv}_{tpl})$.

11.4 THE HELPING PROOF

We now explain the helping proof in detail. We start with the proof of proposition 11.4 as it is the more interesting part. The proof outline is shown in Figure 11.2.

Before we go over the proof details, it is helpful to revisit the rules governing the proof of an atomic triple $\langle x.\ P \rangle\ e\ \langle v.\ Q \rangle$, as discussed in §3.7. Recall that we typically prove an atomic triple by first using the rule LOGATOM-INTRO to obtain ownership of the atomic update token $\mathsf{AU}_{P,Q}(\Phi)$. This token gives us access to the atomic precondition P up to the linearization point. It also serves as an obligation to commit the atomic triple at the linearization point using the rule AU-COMMIT, which demands that P is transformed to Q. This rule then produces the *receipt* $\Phi(v)$, which stands for the precondition of the continuation of the client of the atomic triple and is needed to satisfy the premise of rule LOGATOM-INTRO.

The proof outline in Figure 11.2 uses rule LOGATOM-INTRO to obtain the atomic update token $\mathsf{AU}(\Phi)$ for the atomic triple in proposition 11.4 right at the start of $\overline{\texttt{search}}$ (line 4). Note that we omit the annotation of the pre and postcondition from $\mathsf{AU}(\Phi)$ as it is clear from the context. The proof transfers ownership of $\mathsf{AU}(\Phi)$ from the thread-local context of the $\overline{\texttt{search}}$ thread to the thread-local context of the $\texttt{upsert}$ thread that will linearize the search. We do this via the shared representation predicate $\overline{\mathsf{MCS}}$, or, to be more precise, an invariant that we store in $\overline{\mathsf{MCS}}$. In the proof of $\texttt{upsert}$, when the search is linearized, the associated receipt $\Phi(t')$ is transferred via the shared invariant back to the $\overline{\texttt{search}}$ thread. We next explain the details of this part of the proof along with the definitions of the involved predicates.

Figure 11.3 shows the definition of the representation predicate $\overline{\mathsf{MCS}}$ and the invariant that encodes the helping protocol. The predicate $\overline{\mathsf{MCS}}(\mathsf{Inv}_{tpl}, \mathsf{Prot})(r, t, M)$ contains the predicate $\mathsf{MCS}(r, t, H)$, used in the template-level atomic triples, and then defines M in terms of H via the equality $M = \bar{H}$. All remaining (ghost) resources associated with the data structure are owned by the predicate $\mathsf{Inv}(\mathsf{Inv}_{tpl}, \mathsf{Prot})(r)$. In particular, this predicate stores the ghost resource $\boxed{\bullet H}^{\gamma_s}$ whose type is the authoritative RA over sets $\mathbb{N} \times \mathbb{N}_\perp$. This resource holds the authoritative version of the current upsert history H. Notably, we can then define the predicate $\mathsf{SR}(k, t_0)$ as the fractional resource $\boxed{\circ \{(k, t_0)\}}^{\gamma_s}$, which expresses the auxiliary precondition $(k, t_0) \in H$ needed for search recency.

Inv also stores the abstract template-level invariant $\mathsf{Inv}_{tpl}(r, t, H)$, and the abstract predicate $\mathsf{Prot}(H)$ used for the bookkeeping related to the helping protocol. In addition, Inv states two invariants, $\mathsf{Init}(H)$ and $\mathsf{MaxTS}(t, H)$ that are needed to prove the atomic triple of $\texttt{search}'$.

Note that we add $\mathsf{Inv}(\mathsf{Inv}_{tpl}, \mathsf{Prot})(r)$ to $\overline{\mathsf{MCS}}$ as an Iris invariant (indicated by the box surrounding the predicate). This provides more flexibility when proving that a template operation satisfies its template-level atomic triple. By storing all resources in an invariant, the resources can be accessed in each atomic step, regardless of whether the operation has already passed its linearization point or not.

As Inv is an invariant, we must ensure that it is preserved in each atomic step. However, H and t change with each $\texttt{upsert}$, which means that these values must be existentially quantified

1 $\{\mathsf{SearchSpec}(\mathrm{search}, \mathsf{Inv}_{tpl})\} * \langle t\ M.\ \mathsf{MCS}(\mathsf{Inv}_{tpl}, \mathsf{Prot}_{help})(r, t, M)\rangle$

2 **let** $\overline{\mathrm{search}}$ r k =

3   (* Start application of `logatom-intro` *)

4     $\{\mathsf{AU}(\Phi)\}$

5   **let** tid = NewProph **in**

6   **let** p = NewProph **in**

7     $\{\mathsf{AU}(\Phi) * \mathsf{Proph}(tid, \_) * \mathsf{Proph}(p, vp)\}$

8     $\{\mathsf{AU}(\Phi) * \mathsf{Proph}(tid, \_) * \mathsf{Proph}(p, vp) * \boxed{\mathsf{Inv}(\mathsf{Inv}_{tpl}, \mathsf{Prot}_{help})(r)}\}$

9     $\{\mathsf{AU}(\Phi) * \mathsf{Proph}(tid, \_) * \mathsf{Proph}(p, vp) * \overline{\lceil \circ\ \bar{H}_0 \rceil}^{\gamma_s} * t_0 = \bar{H}_0(k) * \mathsf{SR}(k, t_0)\}$

10   (* Case analysis on $vp < t_0$, $vp = t_0$, $vp > t_0$: only showing $vp > t_0$ *)

11    $\{\mathsf{Proph}(p, vp) * t_0 = \bar{H}_0(k) * \mathsf{SR}(k, t_0) * vp > t_0 * \ldots\}$

12       $\{\ldots * \mathsf{AU}(\Phi) * \mathsf{Proph}(tid, \_)\}$

13       $\{\ldots * \mathsf{Proph}(tid, \_) * \mathsf{Pending}(H_0, k, vp, \Phi) * \lceil \tfrac{1}{2}\ H_0 \rceil^{\gamma_{sy(tid)}} * \lceil \tfrac{1}{2}\ H_0 \rceil^{\gamma_{sy(tid)}} * \mathsf{Tok}\}$

14       $\{\ldots * \lceil \bullet\ R \rceil^{\gamma_r} * tid \notin R * \boxed{\mathsf{State}(tid, k, vp, \Phi, \mathsf{Tok})} * \mathsf{Reg}(tid, H_0, k, vp, \Phi, \mathsf{Tok}) * \mathsf{Tok}\}$

15   (* Ghost update: $\lceil \bullet\ R \rceil^{\gamma_r} \Rightarrow \lceil \bullet\ R \cup \{tid\} \rceil^{\gamma_r}$ *)

16    $\{\mathsf{Proph}(p, vp) * \mathsf{Tok} * \lceil \circ\ \{tid\} \rceil^{\gamma_r} * \boxed{\mathsf{State}(tid, k, vp, \Phi, \mathsf{Tok})} * \mathsf{SR}(k, t_0) * \mathsf{MCS}(r, t, H)\}$

17   **let** t' = search r k **in**

18    $\{\mathsf{Proph}(p, vp) * \mathsf{Tok} * \lceil \circ\ \{tid\} \rceil^{\gamma_r} * \mathsf{MCS}(r, t, H) * (k, t') \in H * t_0 \leq t' * \lceil \{(k, t')\} \rceil^{\gamma_s}\}$

19   Resolve p to t';

20    $\{\mathsf{Tok} * \lceil \circ\ \{tid\} \rceil^{\gamma_r} * vp = t' * \lceil \{(k, vp)\} \rceil^{\gamma_s}\}$

21    $\{\Phi(vp) * vp = t'\}$

22    $\{\Phi(t')\}$

23   t'

24   (* End application of `logatom-intro` *)

25    $\langle t'.\ \mathsf{MCS}(\mathsf{Inv}_{tpl}, \mathsf{Prot}_{help})(r, t, M) * M(k) = t'\rangle$

Figure 11.2: Outline for the proof of proposition 11.4.

by Inv. Nevertheless, we need to ensure that the values t and H exposed in the representation predicate $\mathsf{MCS}(r, t, H)$ of the template-level atomic triples agree with the values stored in the authoritative resources inside Inv. We do this by introducing an additional predicate $\mathsf{MCS}^\bullet(r, t, H)$ that we also add to the invariant. We can think of the predicates $\mathsf{MCS}^\bullet(r, t, H)$ and $\mathsf{MCS}(r, t, H)$ as providing two complementary views at the abstract state of the data structure, one from the perspective of the data structure's implementation, and one from the perspective of the client of the template-level atomic triples. Together, they provide the following important properties:

VIEW-UPD

$$\frac{\mathsf{MCS}^\bullet(r, t, H) * \mathsf{MCS}(r, t, H)}{\mathsf{MCS}^\bullet(r, t', H') * \mathsf{MCS}(r, t', H')}$$

VIEW-SYNC

$$\mathsf{MCS}^\bullet(r, t, H) * \mathsf{MCS}(r, t', H') \vdash t = t' \wedge H = H'$$

$$\overline{\mathsf{MCS}}(\mathsf{Inv}_{tpl}, \mathsf{Prot})(r, t, M) := \exists\, H.\, \mathsf{MCS}(r, t, H) \;*\; M = \bar{H} \;*\; \boxed{\mathsf{Inv}(\mathsf{Inv}_{tpl}, \mathsf{Prot})(r)}$$

$$\mathsf{Inv}(\mathsf{Inv}_{tpl}, \mathsf{Prot})(r) := \exists t\, H.\, \mathsf{MCS}^{\bullet}(r, t, H) \;*\; \overbrace{\left[\,\bullet\, H\,\right]}^{\gamma_s} \;*\; \mathsf{Init}(H) \;*\; \mathsf{MaxTS}(t, H)$$
$$*\; \mathsf{Inv}_{tpl}(t, H) \;*\; \mathsf{Prot}(H)$$

$$\mathsf{Init}(H) := \forall k.\, (k, \bot) \in H$$

$$\mathsf{MaxTS}(t, H) := \forall (k, t') \in H.\, t' < t$$

$$\mathsf{Prot}_{help}(H) := \exists R.\, \overbrace{\left[\,\bullet\, R\,\right]}^{\gamma_r} \;*\; \operatorname*{\text{\Large ✳}}_{tid \in R} \exists k\, vp\, \Phi\, \mathsf{Tok}.\, \mathsf{Reg}(tid, H, k, vp, \Phi, \mathsf{Tok})$$

$$\mathsf{Reg}(tid, H, k, vp, \Phi, \mathsf{Tok}) := \mathsf{Proph}(tid, \_) \;*\; \overbrace{\left[\,\tfrac{1}{2}\, H\,\right]}^{\gamma_{sy(tid)}} \;*\; \boxed{\mathsf{State}(tid, k, vp, \Phi, \mathsf{Tok})}$$

$$\mathsf{State}(tid, k, vp, \Phi, \mathsf{Tok}) := \exists\, H.\, \overbrace{\left[\,\tfrac{1}{2}\, H\,\right]}^{\gamma_{sy(tid)}}$$
$$*\; (\mathsf{Pending}(H, k, vp, \Phi) \vee \mathsf{Done}(H, k, vp, \Phi, \mathsf{Tok}))$$

$$\mathsf{Pending}(H, k, vp, \Phi) := \mathsf{AU}(\Phi) \;*\; (k, vp) \notin H$$

$$\mathsf{Done}(H, k, vp, \Phi, \mathsf{Tok}) := (\Phi(vp) \vee \mathsf{Tok}) \;*\; (k, vp) \in H$$

$$\mathsf{SR}(k, t) := \overbrace{\left[\,\circ\, \{(k, t)\}\,\right]}^{\gamma_s}$$

Figure 11.3: Definition of client-level representation predicate and invariants of helping protocol.

The rule VIEW-UPD says that both predicates are required in order to update the views of the data structure. This occurs in the proof of upsert when both the invariant as well as the precondition of the template-level atomic triple are accessed at the linearization point. The rule VIEW-SYNC ensures that both copies are always in sync. The predicates $\mathsf{MCS}^{\bullet}(r, t, H)$ and $\mathsf{MCS}(r, t, H)$ are defined using a combination of authoritative and exclusive RAs. We omit the precise definitions here as they are unimportant for our discussion.

Let us now return to the proof outline in Figure 11.2. After creating the two prophecies tid and p, the proof uses rule AU-ABORT to obtain access to the precondition $\overline{\mathsf{MCS}}(\mathsf{Inv}_{tpl}, \mathsf{Prot}_{help})(r, t, M)$ and opens it to extract the invariant $\boxed{\mathsf{Inv}(\mathsf{Inv}_{tpl}, \mathsf{Prot}_{help})(r)}$ (line 7). As the invariant is persistent, it can be assumed before each of the remaining steps of the proof, though we must also show that it is preserved by each step. In preparation for the call to search, the proof proceeds on line 8 by opening the invariant $\mathsf{Inv}(\mathsf{Inv}_{tpl}, \mathsf{Prot}_{help})(r)$ and using rule AUTH-SET-SNAP to take a snapshot $\overbrace{\left[\,\circ H_0\,\right]}^{\gamma_s}$ of the authoritative upsert history, whose value at this point we denote by H_0. The proof then lets $t_0 := \bar{H}_0(k)$, which implies $(k, t_0) \in H_0$. This fact and rule AUTH-FRAG-OP then yield $\overbrace{\left[\,\circ\, \{(k, t_0)\}\,\right]}^{\gamma_s} = \mathsf{SR}(k, t_0)$ (line 9).

The proof now case splits on whether the value vp prophesied by p satisfies $t_0 < vp$, $t_0 = vp$, or $t_0 > vp$ (line 10). The last case can be easily discharged as it would imply $t' = vp < t_0$ after resolving the prophecy p on line 19, contradicting the postcondition $t_0 \leq t'$ of search. The case $t_0 = vp$ where the call to search completes without interference from any upserts on k is also straightforward. It implies $M(k) = \bar{H}_0(k) = t_0 = vp = t'$, which allows us to commit the atomic triple right away, i.e., without requiring any help from an upsert thread. We therefore show only the case $t_0 < vp$ in detail, which involves the helping protocol captured by the predicate Prot_{help}.

The predicate Prot_{help} keeps track of the search threads that require helping from upsert threads. The IDs of these threads are stored in an authoritative set R at ghost location γ_r. In the case $t_0 < vp$ of the proof, we must therefore register the thread ID tid with the invariant by replacing R with $R \cup \{tid\}$ using rule AUTH-SET-UPD (line 15). The predicate $\mathsf{Prot}_{help}(H_0)$ needs to be preserved by this ghost update, because it is part of the invariant Inv. This forces the proof to transfer some of its thread-local resources, captured by the predicate $\mathsf{Reg}(tid, H_0, k, vp, \Phi, \mathsf{Tok})$, to the invariant. The lines in the proof preceding the update of R establish that this predicate holds. In particular, the proof transfers the predicate $\mathsf{State}(tid, k, vp, \Phi, \mathsf{Tok})$ to $\mathsf{Prot}_{help}(H_0)$. This predicate keeps track of the current state of thread tid in the helping protocol. Initially, the thread is in state $\mathsf{Pending}(H_0, k, vp, \Phi)$. To establish this predicate, the proof needs to transfer $\mathsf{AU}(\Phi)$ to the invariant, as explained earlier. It also needs to ensure $(k, vp) \notin H_0$, which follows from $\bar{H}_0(k) = t_0 < vp$. Finally, the proof creates a fresh exclusive token Tok that it will later trade in for the receipt $\Phi(vp)$ of the committed atomic triple.

A minor technicality at this point in the proof is that we must also turn the predicate State into an Iris invariant before we add it to $\mathsf{Reg}(tid, H_0, k, vp, \Phi, \mathsf{Tok})$. The reason is that the $\overline{\text{search}}$ proof needs to be able to conclude that the Φ it will receive back from the invariant later after the call to search returns, is the same as the one it registers with the helping protocol before calling search. By turning State into an invariant, the predicate becomes a duplicable resource. This allows the proof to keep one copy of the predicate in its thread-local proof context. An unfortunate side effect of this solution is that, as an invariant, the predicate State cannot expose the upsert history as a parameter. However, the definition of State still depends on the exact value of the upsert history that is stored in the authoritative resource in the invariant. We solve this minor technicality by additionally storing the upsert history at an auxiliary fractional resource at the ghost location $\gamma_{sy(tid)}$. The invariant maintains one such ghost resource for each thread tid that is registered for helping. We split this resource half-way between the predicates Reg and State. This ensures that the upsert history H referenced in State is indeed equal to the authoritative one.

Line 13 shows the proof state after $\mathsf{Pending}(H_0, k, vp, \Phi)$ has been established and the new resource $\lceil \overline{H_0} \rceil^{\gamma_{sy(tid)}}$ has been allocated and split into the two halves. These resources are then assembled into $\mathsf{State}(tid, k, vp, \Phi, \mathsf{Tok})$ and then $\mathsf{Reg}(tid, H_0, k, vp, \Phi, \mathsf{Tok})$, as shown on line 14. Assembling the latter also requires the proof to give up ownership of $\mathsf{Proph}(tid, \_)$. By storing these prophecy resources in $\mathsf{Prot}_{help}(H_0)$ for all threads in R, the proof can use rule PROPH-EXCLUSIVE to conclude $tid \notin R$ before adding tid to R. This is needed to show that the assembled $\mathsf{Reg}(tid, H_0, k, vp, \Phi, \mathsf{Tok})$ can indeed be transferred to the invariant during the ghost update.

After the ghost update of R, we arrive at line 16, at which point we are ready to perform the call to $\mathtt{search}$. After using the invariant and $\mathsf{SR}(k, t_0)$ to obtain the atomic triple for $\mathtt{search}$ from $\mathsf{SearchSpec}(\mathtt{search}, \mathsf{Inv}_{tpl})$, the proof opens the precondition $\overline{\mathsf{MCS}}(r, t, M)$ of $\overline{\mathtt{search}}$ once again and extracts the precondition $\mathsf{MCS}(r, t, H)$ for the template-level atomic triple. Here, H refers to the new value of the upsert history at the linearization point of $\mathtt{search}$. Next, the proof executes the call to $\mathtt{search}$ using the template-level atomic triple, which leaves us with the new proof state on line 18. We can now open the invariant again to obtain a new snapshot $\overline{H'}^{\gamma_s}$ of the upsert history and use rule VIEW-SYNC to conclude that the snapshot value H' is the same as the H in $\mathsf{MCS}(r, t, H)$. Together with $(k, t') \in H$, we can then establish the persistent proposition $\overline{\{(k, t')\}}^{\gamma_s}$. After resolving the prophecy p on line 20, we additionally establish $vp = t'$.

To complete the proof, we now open $\mathsf{Inv}(\mathsf{Inv}_{tpl}, \mathsf{Prot}_{help})$ and use the resource $\overline{\circ\, \{tid\}}^{\gamma_r}$ to conclude that $tid \in R$. That is, we have $\mathsf{Reg}(tid, H, k, vp, \Phi, \mathsf{Tok})$ and can now use $\overline{\{(k, vp)\}}^{\gamma_s}$ and the fractional resources at ghost location $\gamma_{sy(tid)}$ to show that the thread must be in state $\mathsf{Done}(H, k, vp, \Phi, \mathsf{Tok})$. Since the thread still owns the unique token Tok, it can be exchanged for $\Phi(vp)$ in the invariant (line 21). Using $\Phi(vp)$, the proof can then complete the initial application of the rule LOGATOM-INTRO to show the atomic triple of $\overline{\mathtt{search}}$.

Proof of Proposition 11.1.   By comparison, proving the client-level specification of $\mathtt{upsert}$ from its template-level specification is relatively simple. Recall that the goal here is to prove

$$\left\langle t\, M.\, \overline{\mathsf{MCS}}(\mathsf{Inv}_{tpl}, \mathsf{Prot}_{help})(r, t, M) \right\rangle \mathtt{upsert}\ r\ k\ \left\langle \overline{\mathsf{MCS}}(\mathsf{Inv}_{tpl}, \mathsf{Prot}_{help})(r, t+1, M[k \rightarrowtail t]) \right\rangle$$

assuming $\mathsf{UpsertSpec}(\mathtt{upsert}, \mathsf{Inv}_{tpl})$. Let us for a moment assume that we have already established $\mathsf{Updatable}(\mathsf{Prot}_{help})$. We can then use this property to obtain the template-level atomic triple for $\mathtt{upsert}$ from $\mathsf{UpsertSpec}(\mathtt{upsert}, \mathsf{Inv}_{tpl})$. Before the call to $\mathtt{upsert}$, we open the precondition $\overline{\mathsf{MCS}}(\mathsf{Inv}_{tpl}, \mathsf{Prot}_{help})(r, t, M)$ to obtain $\mathsf{MCS}(r, t, H)$. Moreover, we obtain the equality $M = \bar{H}$ and use the property $\mathsf{MaxTS}(t, H)$ from the invariant to conclude that $\bar{H}(k) < t$. Applying the atomic triple gives the postcondition $\mathsf{MCS}(r, t+1, H \cup \{(k, t)\})$. From $\bar{H}(k) < t$ it then follows that we have $H \cup \{(k, t)\} = \bar{H}[k \rightarrowtail t] = M[k \rightarrowtail t]$. This allows us to establish the desired postcondition $\overline{\mathsf{MCS}}(r, t+1, M[k \rightarrowtail t])$.

Finally, we need to show that $\mathsf{Updatable}(\mathsf{Prot}_{help})$ is valid at the point where the template-level atomic triple of $\mathtt{upsert}$ is applied. That is, we must show that we can reestablish $\mathsf{Prot}_{help}(H \cup \{(k, t)\})$, assuming $\mathsf{Prot}_{help}(H)$ is true. In this proof, we make use of the fact that $\mathsf{MaxTS}(t, H)$ holds, which we obtain from the invariant. The important point to note here is that any $\overline{\mathtt{search}}$ thread $tid \in R$ that was in state $\mathsf{Pending}(H, k, t, \Phi)$ and thus waiting to be helped by the upsert thread performing the ghost update of H, cannot remain in this state because

$\mathsf{Pending}(H \cup \{(k, t)\}, k, vp, \Phi)$ is unsatisfiable. The ghost update is, thus, forced to commit the atomic triples of all these $\overline{\texttt{search}}$ threads using their update tokens via rule AU-COMMIT. It can do this because the postcondition of these triples are satisfied after H has been updated. The reasoning here follows similar steps as the proof of the postcondition of upsert above. After committing the atomic triple, the receipt $\Phi(t)$ is transferred to $\mathsf{Prot}_{help}(H \cup \{(k, t)\})$ as part of the new state $\mathsf{Done}(H \cup \{(k, t)\}, k, t, \Phi, \mathsf{Tok})$ for each of these threads.

CHAPTER 12

Verifying the LSM DAG Template

In this chapter, we present a template for multicopy structures that generalizes the LSM (log-structured merge) tree discussed in §9.3 to arbitrary directed acyclic graphs (DAGs). We refer to this template as the LSM DAG template. We prove linearizability of the LSM DAG template by verifying that all operations satisfy the template-level atomic triples (§10.3). The template and proof parameterize over the implementation of the single-copy data structures used at the node-level. Instantiating the template for a specific implementation involves only sequential reasoning about the implementation-specific node-level operations.

The high-level proof builds on the time-ordering invariant discussed in the library analogy for multicopy structures in §9.1. The time-ordering invariant is the observation that the copies of a key are ordered from most recent to least recent as one traverses farther away from the root node. A key challenge we have to address in the proof is how to capture the time-ordering invariant on DAGs of unbounded size while preserving the strategy of decomposing the proof into the concurrency-related aspects and the heap-related aspects. Crucially, the keyset RA introduced in §5.2, which was central to achieving this proof decomposition for verifying single-copy data structures, does not extend to multicopy structures as it critically relies on the fact that every key is present in at most one node of the data structure at a time. Instead, we identify an alternative abstraction of the data structure graph that captures the time-ordering invariant. We then show how to encode this abstraction in separation logic using the flow framework.

12.1 THE LSM DAG TEMPLATE

We split the template into two parts. The first part is a template for search and upsert that works on general multicopy structures, i.e., arbitrary DAGs with locally disjoint edgesets. The second part (discussed in §12.3) is a template for a maintenance operation that generalizes the compaction mechanism found in existing list-based LSM tree implementations to tree-like multicopy structures.

Figure 12.1 shows the code of the template for the core multicopy operations. The operations search and upsert closely follow the high-level description of these operations on the LSM tree (§9.3). The operations are defined in terms of implementation-specific helper functions findNext, addContents, and inContents. The upsert additionally uses readClock and

```
1  let rec traverse n k =              13  let rec upsert r k =
2    lockNode n;                       14    lockNode r;
3    match inContents n k with         15    let t = readClock () in
4    | Some t' -> unlockNode n; t'     16    let res = addContents r k t in
5    | None ->                         17    if res then begin
6      match findNext n k with         18      incrementClock ();
7      | Some n' ->                     19      unlockNode r
8        unlockNode n;                 20    end
9        traverse n' k                 21    else begin
10     | None -> unlockNode n; ⊥       22      unlockNode r;
11                                      23      upsert r k
12 let search r k = traverse r k       24    end
```

Figure 12.1: The general template for multicopy operations search and upsert. The template can be instantiated by providing implementations of helper functions inContents, findNext, and addContents. inContents $n\,k$ returns Some t' if $t' = C_n(k) \neq \bot$, and None otherwise. findNext $n\,k$ returns Some n' if n' is the unique node such that $k \in \mathrm{es}(n, n')$, and None otherwise. addContents $r\,k\,t$ adds the key k with the timestamp t to the contents of r. The return value of addContents is a Boolean which indicates whether the insertion was successful (e.g., if r is full, insertion may fail leaving r's contents unchanged).

incrementClock, auxiliary functions that are ghost code. The ghost functions manipulate the ghost state that keeps track of the clock value, facilitating the proof while having no effect on program behavior.

The search operation calls the recursive function traverse on the root node. traverse $n\,k$ first locks the node n and uses the helper function inContents $n\,k$ to check if a copy of key k is contained in n. If a copy of k is found, then its timestamp t' is returned after unlocking n. Otherwise, traverse uses the helper function findNext to determine the unique successor n' of the given node n and operation key k (i.e., the node n' satisfying $k \in \mathrm{es}(n, n')$). If such a successor n' exists, traverse recurses on n'. Otherwise, traverse concludes that there is no copy of k in the data structure and returns $\bot$. Note that this algorithm uses fine-grained concurrency, as the thread executing the search holds at most one lock at any point (and no locks at the points when traverse is called recursively).

The upsert $r\,k$ operation locks the root node and adds a new version of the key k to the contents of the root node using addContents. In order to add a new copy of k, it must know the current clock value. This is accomplished by using the readClock function. addContents $r\,k\,t$ adds the pair (k, t) to the root node when it succeeds. upsert terminates by incrementing the clock value and unlocking the root node. The addContents function may however fail if the root node is full. In this case upsert calls itself recursively[1].

---

[1]For simplicity of presentation, we assume that a separate maintenance thread flushes the root if it is full to ensure that upserts eventually make progress.

12.2   VERIFYING THE TEMPLATE

We next discuss the correctness proof of the template operations. For each of the operations, we will first focus on the high-level proof ideas and key invariants, then discuss the encoding of the invariants in Iris, before presenting the detailed proof outlines.

12.2.1   HIGH-LEVEL PROOF OUTLINE

Proof of `search`.   We start with the proof of `search`. Recall that search recency is the affirmation that if t_0 is the logical timestamp of k at the point when `search r k` is invoked, then the operation returns $(k, t') \in H$ such that $t' \geq t_0$. Since the timestamp t' (and, in the full implementation, the value) of k retrieved by `search` comes from some node in the structure, we must examine the relationship between the upsert history H of the data structure and the physical contents C_n of the nodes n visited as the search progresses. We do this by identifying the main invariants needed for proving search recency for arbitrary multicopy structures.

We refer to the *spatial ordering* of the copies (k, t) stored in a multicopy structure as the ordering in which these copies are reached when traversing the data structure graph starting from the root node. Our first observation is that the spatial ordering is consistent with the temporal ordering in which the copies have been upserted. We referred to this property as the time-ordering invariant in our library analogy in §9.1: the farther from the root a search is, the older the copies it finds are. Therefore, if a `search r k` traverses the data structure without interference from other threads and returns the first copy of k that it finds, then it is guaranteed to return the logical timestamp of k at the start of the search.

We formalize this observation in terms of the *contents-in-reach* of a node. The contents-in-reach of a node n is the function $C_{ir}(n): \mathbb{K} \to \mathbb{N}_\perp$ defined recursively over the graph of the multicopy structure as follows:

$$C_{ir}(n)(k) := \begin{cases} C_n(k) & \text{if } C_n(k) \neq \perp \\ C_{ir}(n')(k) & \text{else if } \exists n'.\, k \in \text{es}(n, n') \\ \perp & \text{otherwise} \end{cases} \tag{12.1}$$

Note that $C_{ir}(n)$ is well-defined because the graph is acyclic and the edgesets labeling the outgoing edges of every node n are disjoint.

For example, in the multicopy structure depicted in Figure 10.1, we have $C_{ir}(r) = \{k_1 \rightarrowtail 6, k_2 \rightarrowtail 7, k_3 \rightarrowtail 4\}$ and $C_{ir}(n_3) = C_{n_3}$.

The observation that interference-free searches will find the current logical timestamp of their operation key is then captured by the following invariant:

Invariant 1 At every atomic step, the logical contents of the multicopy structure is the contents-in-reach of its root node: $\bar{H} = C_{ir}(r)$.

In order to account for concurrent threads interfering with the search, we prove the condition $t_0 \leq t'$ for the timestamp t' returned by the search. Intuitively, this is true because the contents-in-reach of a node n can be affected only by upserts or maintenance operations, both of which only increase the timestamps associated with every key of any given node: upserts insert new copies into the root node and maintenance operations move recent copies down in the structure, possibly replacing older copies. This observation is formally captured by the following invariant:

Invariant 2 From one atomic step to the next, the contents-in-reach of every node can only increase. That is, for every node n and key k, if $C_{ir}(n)(k) = t$ at some point in time and $C_{ir}(n)(k) = t'$ at any later point in time, then $t \leq t'$.

Finally, in order to prove the condition $(k, t') \in H$ of search recency, we need one additional property:

Invariant 3 At every atomic step, all copies present in the multicopy structure have been upserted at some point in the past. That is, for all nodes n, $C_n \subseteq H$.

Now let us consider an execution of `search` on a operation key k. In addition to the above three general invariants, we need an inductive invariant for the traversal performed by the search: we require as a precondition for `traverse` n k that $C_{ir}(n)(k) \geq t_0$ where t_0 is the logical timestamp of k at the point when `search` was invoked. To see that this property holds initially for the call to `traverse` r k in `search`, let $\bar{H}_0$ be the logical contents at the time point when `search` was invoked. The precondition $SR(k, t_0)$ implies $\bar{H}_0(k) \geq t_0$, which, combined with Invariant 1, implies that we must have had $C_{ir}(r)(k) \geq t_0$ at this point. Since $C_{ir}(r)(k)$ only increases over time because of Invariant 2, we can conclude that $C_{ir}(r)(k) \geq t_0$ when `traverse` is called. We next show that the traversal invariant is maintained by `traverse` and is sufficient to prove search recency.

Consider a call to `traverse` n k such that $C_{ir}(n)(k) \geq t_0$ holds initially. We must show that the call returns t such that $t \geq t_0$ and $(k, t) \in H$. We know that the call to `inContents` on line 3 returns either Some t' such that $t' = C_n(k)$ or None if $C_n(k) = \bot$. Let us first consider the case where `inContents` returns Some t'. In this case, `traverse` returns t' on line 4. By definition of $C_{ir}(n)$, we have $C_{ir}(n)(k) = C_n(k)$. Hence, the precondition $C_{ir}(n)(k) \geq t_0$, together with Invariant 2, implies $t' \geq t_0$. Moreover, Invariant 3 guarantees $(k, t') \in H$.

Now consider the case where `inContents` returns None. Here, $C_n(k) = \bot$, indicating that no copy has been found for k in n. In this case, `traverse` calls `findNext` to obtain the successor node of n and k. In the case where the successor n' exists (line 7), we know that $k \in es(n, n')$ must hold. Hence, by definition of contents-in-reach we must have $C_{ir}(n)(k) = C_{ir}(n')(k)$. From $C_{ir}(n)(k) \geq t_0$ and Invariant 2, we can then conclude $C_{ir}(n')(k) \geq t_0$, i.e., that the precondition for the recursive call to `traverse` on line 9 is satisfied and search recency follows by induction.

On the other hand, if n does not have any next node, then `traverse` returns $\bot$ (line 10), indicating that k has not yet been upserted at all so far (i.e., has never appeared in the structure). In this case, by definition of contents-in-reach we must have $C_{ir}(n)(k) = \bot$. Invariant 2 then guarantees $\perp \geqslant t_0$. By definition of the ordering on timestamps extended with $\perp$, we also have $\perp \leqslant t_0$, which in turn implies $t_0 = \perp$. Hence, search recency holds trivially in the $\perp$ case.

Proof of upsert. In order to prove the logically atomic specification (11.1) of upsert, we must identify an atomic step where the clock t is incremented and the upsert history H is updated. Intuitively, this atomic step is when the global clock is incremented (line 19 in Figure 12.1) after addContents succeeds. Note that in this case addContents changes the contents of the root node from C_r to $C'_r = C_r[k \rightarrowtail t]$. Hence, in the proof we need to update the ghost state for the upsert history from H to $H' = H \cup \{(k, t)\}$, reflecting that a new copy of k has been upserted. It then remains to show that the three key high-level invariants of multicopy structures identified above are preserved by these updates.

First, observe that Invariant 3, which states $\forall n.\ C_n \subseteq H$, is trivially maintained: only C_r is affected by the upsert and the new copy (k, t) is included in H'. Similarly, we can easily show that Invariant 2 is maintained: $C_{ir}(n)$ remains the same for all nodes $n \neq r$ and for the root node it increases, provided Invariant 1 is also maintained.

Thus, the interesting case is Invariant 1. Proving that this invariant is maintained amounts to showing that $\bar{H}'(k) = t$. This step critically relies on the following additional observation:

Invariant 4 At every atomic step, all timestamps in H are smaller than the current time of the global clock t.

This invariant implies that $\bar{H}'(k) = \max(\bar{H}(k), t) = t$, which proves the desired property. We note that Invariant 4 is maintained because the global clock is incremented when H is updated to H', and, as we describe below, while r is locked.

In the proof of Invariant 1 we have silently assumed that the timestamp t, which was obtained by reading the global clock at line 15, is still the value of the global clock at the linearization point when the clock is incremented at line 18. This step in the proof relies on the observation that only upsert changes the global clock and it does so only while the clock is protected by the root node's lock. Hence, for lock-based implementations of multicopy structures, we additionally require the following invariant:

Invariant 5 From one atomic step to the next, if a thread holds the lock on the root node, no other thread will change the value of the global clock.

In the remainder of this section, we discuss how to formalize this high-level proof outline in Iris.

12.2.2 IRIS INVARIANT

The Iris proof will need to capture the key invariants identified in the proof outline given above in terms of appropriate ghost state constructions. We start by addressing the key technical issue that arises when formalizing the above proof in a separation logic like Iris: contents-in-reach is a recursive function defined over an arbitrary DAG of unbounded size. This makes it difficult to obtain a simple local proof that involves reasoning only about the bounded number of modified nodes in the graph. The recursive and global nature of contents-in-reach mean that modifying even a single edge in the graph can potentially change the contents-in-reach of an unbounded number of nodes (for example, deleting an edge (n_1, n_2) can change $C_{ir}(n)$ for all n that can reach n_1). A straightforward attempt to prove that a template algorithm preserves Invariant 2 would thus need to reason about the entire graph after every modification (for example, by performing an explicit induction over the full graph). We solve this challenge using the flow framework.

Encoding Contents-in-Reach using Flows. Let us revisit the recursive definition of contents-in-reach given in Equation (12.1). The computation of contents-in-reach proceeds bottom-up in the multicopy structure graph starting from the leaves and continues recursively toward the root node. That is, the computation proceeds *backward* with respect to the direction of the graph's edges. This makes a direct encoding of contents-in-reach in terms of a flow difficult because the flow equation (FlowEqn) describes computations that proceed in the forward direction.

We side-step this problem by tracking auxiliary ghost information in the data structure invariant for each node n in the form of a function $Q_n \colon \mathbb{K} \to \mathbb{N}_\perp$. If these ghost values satisfy

$$Q_n = \lambda k. \begin{cases} C_{ir}(n')(k) & \text{if } \exists n'. \ k \in \mathsf{es}(n, n') \\ \perp & \text{otherwise} \end{cases} \tag{12.2}$$

and we additionally define

$$B_n := \lambda k. \ (C_n(k) \neq \perp \ ? \ C_n(k) : Q_n(k))$$

then $C_{ir}(n) = B_n$. The idea is that each node stores Q_n so that node-local invariants can use it to talk about $C_{ir}(n)$. We then use a flow to propagate the purported values Q_n forward in the graph to ensure that they indeed satisfy the condition (12.2). Note that while an upsert or maintenance operations on n may change B_n, it preserves Q_n. That is, operations do not affect the contents-in-reach of downstream nodes, allowing us to reason locally about the modification of the contents of n.

In what follows, let us fix a multicopy structure over nodes N and some valuations of the functions Q_n. The flow domain M for our encoding of contents-in-reach consists of multisets of key-timestamp pairs $M := \mathbb{K} \times \mathbb{N}_\perp \to \mathbb{N}$ with multiset union as the monoid operation. The edge function induced by the multicopy structure is defined as follows:

$$e(n, n')(\_) := \chi(\{(k, Q_n(k)) \mid k \in \mathsf{es}(n, n')\}) \tag{12.3}$$

Here, χ takes a set to its corresponding multiset. Additionally, we let the function *in* map every node to the empty multiset. With the definitions of e and *in* in place, there exists a unique flow *fl* that satisfies (FlowEqn). Now, if every node n in the resulting flow graph satisfies the following two predicates

$$\phi_1(n) := \forall k.\ Q_n(k) = \bot \vee (\exists n'.\ k \in es(n, n')) \tag{12.4}$$

$$\phi_2(n) := \forall k\ t.\ fl(n)(k, t) > 0 \rightarrow B_n(k) = t \tag{12.5}$$

then $B_n = C_{ir}(n)$. Note that the predicates ϕ_1 and ϕ_2 depend only on n's own flow and its local ghost state (i.e., Q_n, C_n and the outgoing edgesets $es(n, \_)$).

The following lemma states the correctness of this encoding.

Lemma 12.1   *If $\phi_1(n, B_n, C_n, I_n)$ and $\phi_2(n, B_n, I_n)$ hold for all nodes n in the flow graph obtained from a multicopy structure as described above, then $B_n = C_{ir}(n)$.*

Proof. The proof proceeds by induction over the inverse topological order of the multicopy structure graph. Let k be a key and n a node in the graph. We show $B_n(k) = C_{ir}(n)(k)$, provided $B_{n'} = C_{ir}(n')$ holds for all successors n' of n in the topological order. We have to consider three cases according to the definition of $C_{ir}(n)$ (Equation (12.1)).

First, suppose $C_n(k) \neq \bot$. In this case, we have $C_{ir}(n)(k) = C_n(k)$, as well as $B_n(k) = C_n(k)$ by definition of B_n.

Next, suppose $C_n(k) = \bot$ and $k \in es(n, n')$ for some n'. Then, by definition of $C_{ir}(n)(k)$ and induction hypothesis on n', we get $C_{ir}(n)(k) = C_{ir}(n')(k) = B_{n'}(k)$. Additionally, Equation (12.3) implies $I_n.out(n')(k, Q_n(k)) > 0$. Since we have a valid flow graph, the interfaces I_n and $I_{n'}$ must compose, which implies $I_{n'}.in(n')(k, Q_n(k)) > 0$. It follows from $\phi_2(n', B_{n'}, I_{n'})$ that we then must have $B_{n'}(k) = Q_n(k) = B_n(k)$, where the second equality follows from the definition of B_n. Hence, we conclude that $C_{ir}(n)(k) = B_n(k)$.

Finally, consider the case where $C_n(k) = \bot$ and for all nodes n', $k \notin es(n, n')$. In this case, $\phi_1(n, B_n, C_n, I_n)$ implies that $Q_n(k) = \bot$. Thus, we again conclude $C_{ir}(n)(k) = \bot = Q_n(k) = B_n(k)$. $\square$

Encoding the Invariants in Iris. We can now define the template-specific invariant $\mathsf{Inv}_{tpl}(r, t, H)$ for the LSM DAG template, which is assumed by the representation predicate $\mathsf{MCS}(\mathsf{Inv}_{tpl}, \mathsf{Prot})(r, t, H)$ defined in Figure 11.3. We denote this invariant by Inv_{LSM} (Figure 12.2).

The parameters t and H are to be interpreted as the current clock value and the current upset history. The existentially quantified variable I is the current global flow interface used for encoding the contents-in-reach flow. The invariant consists of two parts: a *global* part described by the predicate G and a *local* part that holds for every node in the data structure. The set of nodes N of the data structure graph is implicitly captured by the domain $\mathsf{dom}(I)$ of the flow interface I.

As discussed in §4.4, the predicate $\mathsf{L}(b, n, R_n)$ captures the abstract state of n's lock and is used to specify the protocol providing exclusive access to the resource R_n protected by the lock

$$\mathsf{Inv}_{LSM}(r, t, H) := \exists\, I.$$
$$\mathsf{G}(r, t, I)$$
$$* \mathop{\text{\Large{$*$}}}_{n \in \mathsf{dom}(I)} \exists b_n\, C_n\, Q_n.\ \mathsf{L}(b_n, n, \mathsf{N_L}(r, n, C_n, Q_n))$$
$$* \mathsf{N_S}(r, n, C_n, Q_n, H)$$

where $\quad \mathsf{G}(r, t, I) := \boxed{\tfrac{1}{2} \bullet t}^{\gamma_t}$
$$* \boxed{\bullet\, I}^{\gamma_I} * \boxed{\bullet\, \mathsf{dom}(I)}^{\gamma_f}$$
$$* I.in = \lambda_0 * \mathsf{InFP}(r)$$

$$\mathsf{InFP}(n) := \boxed{\circ\, \{n\}}^{\gamma_f}$$

$$\mathsf{N_L}(r, n, C_n, Q_n) := \exists es\, t.\, \mathsf{Node}(r, n, es, C_n)$$
$$* \boxed{\tfrac{1}{2}es}^{\gamma_{e(n)}} * \boxed{\tfrac{1}{2}C_n}^{\gamma_{c(n)}} * \boxed{\tfrac{1}{2}Q_n}^{\gamma_{q(n)}} * \boxed{\circ\, C_n}^{\gamma_s}$$
$$* \left(n = r\,?\, \boxed{\tfrac{1}{2} \bullet t}^{\gamma_t} : \mathsf{True}\right)$$

$$\mathsf{N_S}(r, n, C_n, Q_n, H) := \exists es\, I_n.$$
$$* \boxed{\tfrac{1}{2}es}^{\gamma_{e(n)}} * \boxed{\tfrac{1}{2}C_n}^{\gamma_{c(n)}} * \boxed{\tfrac{1}{2}Q_n}^{\gamma_{q(n)}}$$
$$* \boxed{\circ\, I_n}^{\gamma_I} * \mathsf{dom}(I_n) = \{n\} * \mathsf{InFP}(n) * \mathsf{closed}(es)$$
$$* I_n.out = \lambda n'.\ \chi(\{(k, Q_n(k)) \mid k \in es(n')\})$$
$$* \left(n = r\,?\, B_n = \bar{H} * I_n.in = \lambda_0 : \mathsf{True}\right)$$
$$* \mathop{\text{\Large{$*$}}}_{k \in \mathbb{K}} \boxed{\bullet\, B_n(k)}^{\gamma_{cir(n)(k)}}$$
$$* \phi_1(n, Q_n, es) * \phi_2(n, C_n, Q_n, I_n)$$

$$\mathsf{closed}(es) := \forall n'.\ es(n') \neq \emptyset \rightarrow \mathsf{InFP}(n')$$
$$B_n := \lambda k.\ (C_n(k) \neq \bot\,?\, C_n(k) : Q_n(k))$$
$$\phi_1(n, Q_n, es) := \forall k.\ Q_n(k) = \bot \vee (\exists n'.\, k \in es(n'))$$
$$\phi_2(n, C_n, Q_n, I_n) := \forall k\, t.\ I_n.in(n)(k, t) > 0 \rightarrow B_n(k) = t$$

Figure 12.2: The Iris invariant for the LSM DAG template.

via the helper functions `lockNode` and `unlockNode`. The Boolean b indicates whether the lock is (un)locked.

For the most part, Inv_{LSM} uses similar RA constructions as the invariants of the single-copy structure templates discussed in §8, including authoritative sets and flow interfaces as well as various fractional RAs. In addition, it uses the authoritative RA of natural numbers with maximum as the underlying monoid operation (referred to from now on as the authoritative maxnat RA). This RA guarantees the following properties:

AUTH-MAXNAT-VALID
$$\frac{\mathcal{V}(\bullet m \cdot \circ n)}{m \geq n}$$

AUTH-MAXNAT-UPD
$$\frac{m \leq n}{\bullet m \leadsto \bullet n}$$

AUTH-MAXNAT-SNAP
$$\bullet m \leadsto \bullet m \cdot \circ m$$

The ghost resource $\bullet\, m$ signifies that the current value is m and the ghost resource $\circ\, n$ expresses that n is a lower bound on the current value. That is, $\bullet m \cdot \circ n$ can be valid only if $m \geq n$, as captured by rule AUTH-MAXNAT-VALID. Consequently, the only frame-preserving update permitted by this RA is to replace the current value m by any larger value (rule AUTH-MAXNAT-UPD). Finally, rule AUTH-MAXNAT-SNAP allows us to take a snapshot of the current value m and remember it as a lower bound for a value of the resource that is observed at a later point in the proof.

We discuss each part of Figure 12.2 in detail, starting with the predicate $\mathsf{G}(r, t, I)$:

- We track the current value t of the global clock in a fractional authoritative maxnat RA at ghost location γ_t. The RA ensures that the clock value can only increase. To capture Invariant 5, we split this resource half-way between the invariant and the predicate $\mathsf{N_L}$ for the root node, which is protected by the lock of the root node.

- We use an authoritative RA of flow interfaces at ghost location γ_I to keep track of the global interface I, which is composed of singleton interfaces I_n for each node $n \in \mathsf{dom}(I)$. The I_n are tied to the implementation-specific physical representation of the individual nodes via the predicates $\mathsf{N_L}$ and $\mathsf{N_S}$ as explained below.

- The ghost resource $\lceil \bullet\, \mathsf{dom}(I) \rceil^{\gamma_f}$ keeps track of the footprint of the data structure using the authoritative RA of sets of nodes. We use this resource to maintain the invariant of $\texttt{traverse}$ that the currently visited node n remains part of the data structure while n is unlocked.

- We require that the global flow interface I has no inflow ($I.in = \lambda_0$), as required for our encoding of contents-in-reach. That is, λ_0 maps all nodes to the empty multiset, the unit 0 of the flow domain.

- The condition $\mathsf{InFP}(r)$ guarantees that r is always in the domain of the data structure.

The resources for every node n are split between the two predicates $\mathsf{N_L}(r, n, C_n, Q_n)$ and $\mathsf{N_S}(r, n, C_n, Q_n, H)$. The latter is always owned by the invariant whereas the former is protected by n's lock and transferred between the invariant and the thread's local state upon locking the node and vice versa upon unlocking, as usual. We next discuss $\mathsf{N_L}(r, n, C_n, Q_n)$:

- The first conjunct is the implementation-specific node predicate $\mathsf{Node}(r, n, es, C_n)$. For each specific implementation of the template, this predicate must tie the physical representation of the node n to its contents C_n and a function $es : \mathfrak{N} \to \wp(\mathbb{K})$ which captures the edgesets of n's outgoing edges. Our template proof is parametric in the definition of Node and depends only on the following two assumptions that each implementation used to instantiate the template must satisfy. First, we require that Node is not duplicable:

$$\mathsf{Node}(r, n, es, C_n) * \mathsf{Node}(r', n, es', C'_n) \vdash \mathsf{False}$$

Moreover, Node must guarantee disjoint edgesets:

$$\mathsf{Node}(r, n, es, C_n) \vdash \forall n_1\, n_2.\, n_1 = n_2 \vee es(n_1) \cap es(n_2) = \emptyset$$

- The fractional resources at ghost locations $\gamma_{e(n)}$, $\gamma_{c(n)}$, and $\gamma_{q(n)}$ ensure that the predicates $\mathsf{N_L}$ and $\mathsf{N_S}$ agree on the values of es, C_n, and Q_n even when n is locked.

- The ghost resource $\lceil \circ\, C_n \rceil^{\gamma_s}$ when combined with $\lceil \bullet\, H \rceil^{\gamma_s}$ implies $C_n \subseteq H$, which captures Invariant 3.

- The final conjunct of $\mathsf{N_L}$ guarantees sole ownership of the global clock by a thread holding the lock on the root node.

Moving on to $\mathsf{N_S}(r, n, C_n, Q_n, H)$, this predicate contains those resources of n that are available to all threads at all times:

- The resource $\lceil \circ I_n \rceil^{\gamma_I}$ guarantees that all the singleton interfaces I_n compose to the global interface $\overline{I}$, thus, satisfying the flow equation. Similarly, the predicate $\mathsf{InFP}(n)$ guarantees that n remains in the data structure at all times. The predicate $\mathsf{closed}(es)$ ensures that the outgoing edges of n point to nodes which are again in the data structure. Together with the condition $\mathsf{InFP}(r)$, this guarantees that all nodes reachable from r must be in $\mathsf{dom}(I)$.

- The next conjunct of $\mathsf{N_S}$ defines the outflow of the singleton interface I_n according to Equation (12.3) of our flow encoding of contents-in-reach. Note that, even though I_n is not shared with the predicate $\mathsf{N_L}$, only its inflow can change when n is locked, because the outflow of the interface is determined by Q_n and es, both of which are protected by n's lock.

- The constraint $B_n = \overline{H}$ holds if $r = n$ and implies Invariant 1. We further require here that the interface of the root node has no inflow ($I_n.in = \lambda_0$), a property that we need in order to prove that upsert maintains the flow-related invariants. Moreover, we use for every key k an authoritative maxnat resource at ghost location $\gamma_{cir(n)(k)}$ to capture Invariant 2.

- The last two conjuncts of $\mathsf{N_S}$ complete our encoding of contents-in-reach and ensure that we must indeed have $B_n = C_{ir}(n)$ at all atomic steps.

Finally, we note that Invariant 4 is already captured by the predicate $\mathsf{MaxTS}(t, H)$ included in the full multicopy structure invariant $\mathsf{Inv}(\mathsf{Inv}_{LSM}, \mathsf{Prot})(r)$.

1 $\langle b\ R.\ \mathsf{L}(b,n,R)\rangle$ lockNode n $\langle \mathsf{L}(\mathit{true},n,R) * R\rangle$
2 $\langle R.\ \mathsf{L}(\mathit{true},n,R) * R\rangle$ unlockNode n $\langle \mathsf{L}(\mathit{false},n,R)\rangle$
3
4 $\{\mathsf{Node}(r,n,es,C_n)\}$
5 inContents $n\ k$
6 $\{v.\ \mathsf{Node}(r,n,es,C_n) * (v = \mathsf{Some}(t) * t = C_n(k) \neq \bot \vee v = \mathsf{None} * C_n(k) = \bot)\}$
7
8 $\{\mathsf{Node}(r,n,es,C_n)\}$
9 findNext $n\ k$
10 $\{v.\ \mathsf{Node}(r,n,es,C_n) * (v = \mathsf{Some}(n') * k \in es(n') \vee v = \mathsf{None} * \forall n'.\ k \notin es(n'))\}$

Figure 12.3: Specifications of helper functions used by search.

12.2.3   DETAILED PROOF OUTLINE

We now have all the ingredients to walk through the Iris proofs of the template operations.

Proof of search.   We start with the specification of the implementation-specific helper functions assumed by search as well as the lock module used in the proof. They are provided in Figure 12.3. The function findNext assumes ownership of the resources $\mathsf{Node}(r,n,es,C_n)$ associated with the locked node n and returns a successor node n' of n such that k is in the edgeset of (n,n') if such a node exists.

Figure 12.4 provides the outline of the proof of search. The intermediate assertions shown throughout the proof represent the relevant information from the proof context at the corresponding point. By convention, all the newly introduced variables are existentially quantified. Note that the condition $\mathsf{SR}(k,t_0)$ is persistent and, hence, holds throughout the proof. Moreover, the invariant $\boxed{\mathsf{Inv}(\mathsf{Inv}_{LSM},\mathsf{Prot})(r)}$ is maintained throughout the proof since search does not modify any shared resources. We therefore do not include these resources explicitly in the intermediate assertions.

As most of the actual work is done by the recursive function traverse, we start with its atomic specification:

$$\frac{\boxed{\mathsf{Inv}(\mathsf{Inv}_{LSM},\mathsf{Prot})(r)} \mathrel{-\!*} \mathsf{InFP}(n) \mathrel{-\!*} \ulcorner \circ t_1 \urcorner^{\gamma_{cir(n)(k)}} \mathrel{-\!*} t_0 \leqslant t_1 \mathrel{-\!*}}{\langle t\ H.\ \mathsf{MCS}(r,t,H)\rangle\ \mathtt{traverse}\ n\ k\ \langle t'.\ \mathsf{MCS}(r,t,H) * t_0 \leqslant t' * (k,t') \in H\rangle}$$

Recall the traversal invariant $t_0 \leqslant C_{ir}(n)(k)$ that we used in our informal proof of search recency. The ghost resource $\ulcorner \circ t_1 \urcorner^{\gamma_{cir(n)(k)}}$, together with $t_0 \leqslant t_1$ in the precondition of the above specification precisely capture this invariant. In addition, traverse assumes the invariant $\mathsf{Inv}(\mathsf{Inv}_{LSM},\mathsf{Prot})(r)$ and requires that n must be a node in the graph, expressed by the predicate $\mathsf{InFP}(n)$. The operation then guarantees to return t' such that search recency holds.

Let us for now assume that traverse satisfies the above specification and focus on the proof of search. That is, we prove proposition 11.2 where we instantiate Inv_{tpl} with Inv_{LSM}.

$1\quad \left\{\boxed{\mathsf{Inv}(\mathsf{Inv}_{LSM},\mathsf{Prot})(r)} * \mathsf{SR}(k,t_0)\right\} * \left\langle t\ H.\ \mathsf{MCS}(r,t,H)\right\rangle$

$2\ \mathbf{let}\ \mathrm{search}\ r\ k\ =$

$3\quad \left\{\mathsf{InFP}(r) * \bar{H}_1(k) = B_r(k) * \overline{\lceil \circ\ B_r(k)\ \rceil}^{\gamma_{cir(r)(k)}} * t_0 \leqslant \bar{H}_1(k)\right\}$

$4\quad \left\{\mathsf{InFP}(r) * \overline{\lceil \circ\ t_1\ \rceil}^{\gamma_{cir(r)(k)}} * t_0 \leqslant t_1\right\} * \left\langle t\ H.\ \mathsf{MCS}(r,t,H)\right\rangle$

$5\quad \mathrm{traverse}\ r\ k$

$6\ \left\langle t'.\ \mathsf{MCS}(r,t,H) * t_0 \leqslant t' * (k,t') \in H\right\rangle$

7

$8\quad \left\{\boxed{\mathsf{Inv}(\mathsf{Inv}_{LSM},\mathsf{Prot})(r)} * \mathsf{InFP}(n) * \overline{\lceil \circ\ t_1\ \rceil}^{\gamma_{cir(n)(k)}} * t_0 \leqslant t_1\right\} * \left\langle t\ H.\ \mathsf{MCS}(r,t,H)\right\rangle$

$9\ \mathbf{let}\ \mathbf{rec}\ \mathrm{traverse}\ n\ k\ =$

$10\quad \mathrm{lockNode}\ n;$

$11\quad \left\{\mathsf{N_L}(r,n,C_n,Q_n) * \overline{\lceil \circ\ B_n(k)\ \rceil}^{\gamma_{cir(n)(k)}} * t_0 \leqslant B_n(k)\right\}$

$12\quad \mathbf{match}\ \mathrm{inContents}\ n\ k\ \mathbf{with}$

$13\quad \mid\ \mathrm{Some}\ t'\ \rightarrow$

$14\quad \left\{\mathsf{N_L}(r,n,C_n,Q_n) * t_0 \leqslant B_n(k) * t' = C_n(k) \neq \bot\right\}$

$15\quad (*\ \mathrm{Linearization\ point}\ *)$

$16\quad \left\{\mathsf{N_L}(r,n,C_n,Q_n) * t_0 \leqslant C_n(k) * t' = C_n(k)\right\}$

$17\quad \left\{\mathsf{N_L}(r,n,C_n,Q_n) * (k,t') \in H * t_0 \leqslant t' * \mathsf{MCS}(r,t,H)\right\}$

$18\quad \mathrm{unlockNode}\ n;\ t'$

$19\quad \left\langle t'.\ \mathsf{MCS}(r,t,H) * (k,t') \in H * t_0 \leqslant t'\right\rangle$

$20\quad \mid\ \mathrm{None}\ \rightarrow$

$21\quad \left\{\mathsf{N_L}(r,n,C_n,Q_n) * t_0 \leqslant B_n(k) * C_n(k) = \bot\right\}$

$22\quad \mathbf{match}\ \mathrm{findNext}\ n\ k\ \mathbf{with}$

$23\quad \mid\ \mathrm{Some}\ n'\ \rightarrow$

$24\quad \left\{\begin{array}{l}\mathsf{Node}(r,n,es,C_n) * \cdots * t_0 \leqslant B_n(k) * C_n(k) = \bot \\ \qquad\qquad\qquad\qquad\qquad\quad * k \in es(n')\end{array}\right\}$

$25\quad \left\{\mathsf{N_L}(r,n,C_n,Q_n) * \mathsf{InFP}(n') * \overline{\lceil \circ\ t_1\ \rceil}^{\gamma_{cir(n')(k)}} * t_0 \leqslant t_1\right\}$

$26\quad \mathrm{unlockNode}\ n;$

$27\quad \left\{\mathsf{InFP}(n') * \overline{\lceil \circ\ t_1\ \rceil}^{\gamma_{cir(n')(k)}} * t_0 \leqslant t_1\right\} * \left\langle t\ H.\ \mathsf{MCS}(r,t,H)\right\rangle$

$28\quad \mathrm{traverse}\ n'\ k$

$29\quad \left\langle t'.\ \mathsf{MCS}(r,t,H) * (k,t') \in H * t_0 \leqslant t'\right\rangle$

$30\quad \mid\ \mathrm{None}\ \rightarrow$

$31\quad \left\{\begin{array}{l}\mathsf{Node}(r,n,es,C_n) * \cdots * t_0 \leqslant B_n(k) * C_n(k) = \bot \\ \qquad\qquad\qquad\qquad\qquad\quad * \forall n'.\ k \notin es(n')\end{array}\right\}$

$32\quad \left\{\mathsf{N_L}(r,n,C_n,Q_n) * t_0 \leqslant B_n(k) * B_n(k) = C_n(k) = \bot\right\}$

$33\quad (*\ \mathrm{Linearization\ point}\ *)$

$34\quad \left\{\mathsf{N_L}(r,n,C_n,Q_n) * t_0 = B_n(k) = C_n = \bot * \mathsf{MCS}(r,t,H)\right\}$

$35\quad \left\{\mathsf{N_L}(r,n,C_n,Q_n) * (k,\bot) \in H * t_0 = \bot * \mathsf{MCS}(r,t,H)\right\}$

$36\quad \mathrm{unlockNode}\ n;\ \bot$

$37\ \left\langle t'.\ \mathsf{MCS}(r,t,H) * (k,t') \in H * t_0 \leqslant t'\right\rangle$

Figure 12.4: Proof of search.

The precondition of search r k assumes the invariant $\mathsf{Inv}(\mathsf{Inv}_{LSM}, \mathsf{Prot})(r)$ and the predicate $\mathsf{SR}(k, t_0)$ (line 1). We must therefore use these to establish the precondition for the call traverse n k on line 5. To this end, we open the invariant from which we can directly obtain $\mathsf{InFP}(r)$. Next, we unfold the definition of $\mathsf{SR}(k, t_0)$ to obtain $\boxed{\circ\ \{(k, t_0)\}}^{\gamma_s}$. Snapshotting $\boxed{\bullet\ H_1}^{\gamma_s}$ in the invariant for the current upsert history H_1, we can conclude $(k, t_0) \in H_1$ and therefore $t_0 \leq \bar{H}_1(k)$. From $\mathsf{N_S}(r, r, C_r, Q_r, H_1)$ in the invariant we can further deduce $\bar{H}_1(k) = B_r(k)$ and $\boxed{\circ\ B_r(k)}^{\gamma_{cir(r)(k)}}$ (line 3). By substituting both $\bar{H}_1(k)$ and $B_r(k)$ with a fresh existentially quantified variable t_1 we obtain the precondition of traverse (line 4). We now commit the atomic triple of search on the call to traverse and immediately obtain the desired postcondition.

Proof of traverse.   Finally, we prove the assumed specification of traverse. The proof starts at line 8. The thread first locks node n which yields ownership of the predicate $\mathsf{N_L}(r, n, C_n, Q_n)$. At this point, we also open the invariant to take a fresh snapshot of the resource $\boxed{\bullet\ B_n(k)}^{\gamma_{cir(n)(k)}}$ to conclude $t_0 \leq B_n(k)$ from the precondition of traverse (line 11). Next, the thread executes inContents n k. The precondition of this call, i.e., $\mathsf{Node}(r, n, es, C_n)$, is available to us as part of the predicate $\mathsf{N_L}(r, n, C_n, Q_n)$. Depending on the return value v of inContents we end up with two subcases.

In the case where $v = \mathsf{Some}(t')$, we know $t' = C_n(k) \neq \bot$ and continue on line 14. The call to unlockNode on line 18 will be the linearization point of this case. To obtain the desired postcondition of the atomic triple, we first retrieve $\mathsf{N_S}(r, n, C_n, B_n, H)$ from the invariant and open its definition. From $C_n(k) \neq \bot$ and the definition of B_n we first obtain $B_n(k) = C_n(k)$. This leaves us with the proof context on line 16. Now we access the precondition $\mathsf{MCS}(r, t, H)$ of the atomic triple and sync it with the view of H and t in the invariant. Moreover, we use the resource $\boxed{\circ\ C_n}^{\gamma_s}$ to infer $C_n \subseteq H$, which then implies $(k, t') \in H$ (line 17). The call to unlockNode returns $\mathsf{N_L}(r, n, C_n, Q_n)$ to the invariant and commits the atomic triple, which concludes this case.

For the second case where the return value of inContents is $v = \mathsf{None}$, we have $C_n(k) = \bot$ and the thread calls findNext. Here, we unfold $\mathsf{N_L}(r, n, C_n, Q_n)$ to retrieve $\mathsf{Node}(r, n, es, C_n)$, which is needed to satisfy the precondition of findNext. We then end up again with two subcases: one where there is a successor node n' such that $k \in es(n')$, and the other where no such node exists. Let us consider the first subcase, which is captured by the proof context on line 24. Now, before the thread unlocks n, we have to reestablish the precondition of traverse for the recursive call on line 28. To do this, we first open the invariant and retrieve $\mathsf{N_S}(r, n, C_n, Q_n, H)$. From predicate $\mathsf{closed}(es)$, we obtain the resource $\mathsf{InFP}(n')$ because $es(n') \neq \emptyset$. $\mathsf{InFP}(n')$ is the first piece for the precondition of traverse. To obtain the remaining pieces, we must retrieve $\mathsf{N_S}(r, n', C_{n'}, Q'_n, H)$ from the invariant. This is possible because we can infer $n' \in \mathsf{dom}(I)$ using $\mathsf{InFP}(n')$.

We first observe that $C_n(k) = \bot$ implies $B_n(k) = Q_n(k)$. Together with $k \in es(n')$, this gives us

$$I_n.out(n')(k, B_n(k)) > 0.$$

From the fact that I_n and $I_{n'}$ compose, we can conclude $I_{n'}.in(n')(k, B_n(k)) > 0$. It then follows from the constraint $\phi_2(n', C_n, Q_n, I_{n'})$ that $B_{n'}(k) = B_n(k)$. We can further take a fresh snapshot of the resource $\overline{\bullet\ B_{n'}(k)}^{\gamma_{cir(n')(k)}}$ to obtain our final missing piece $\overline{\circ\ B_{n'}(k)}^{\gamma_{cir(n')(k)}}$ for the precondition of traverse. Then substituting both $B_n(k)$ and $B_{n'}(k)$ by a fresh variable t_1 and folding predicate $N_L(r, n, C_n, Q_n)$ we arrive at line 25. Unlocking n transfers ownership of $N_L(r, n, C_n, Q_n)$ back to the invariant. The resulting proof context satisfies the precondition of the recursive call to traverse, which we use to commit the atomic triple by applying the specification of traverse inductively.

We are left with the last subcase where n has no successor for k (line 31). Here, we proceed similarly to the first case above: we conclude $B_n(k) = Q_n(k)$ from $C_n(k) = \bot$ and use $\forall n'. k \notin es(n')$ and $\phi_1(n, Q_n, es)$ to conclude that $Q_n(k) = B_n(k) = C_n(k) = \bot$. After folding predicate $N_L(r, n, C_n, Q_n)$ we arrive on line 32. The linearization point in this case is at the point when n is unlocked, so we access the precondition $MCS(r, t, H)$ of the atomic triple and sync it with the view of H and t in the invariant. Again, using the resource $\overline{\circ\ C_n}^{\gamma_s}$ we infer $C_n \subseteq H$ to conclude $(k, \bot) \in H$. The call to unlockNode returns $N_L(r, n, C_n, Q_n)$ to the invariant and commits the atomic triple. The postcondition follows for $t' = \bot$, which is the return value in this case.

This completes the proof of search.

Proof of upsert. We now focus on proving the correctness of the upsert operation. That is, we prove proposition 11.1 where we instantiate Inv_{tpl} with Inv_{LSM}. Figure 12.5 shows the proof outline. The first three lines provide the specification of the implementation-specific helper function addContents that we assume in the proof. The specification simply says that when the function succeeds, the copy of k is updated to t in C_r. If the function fails, then no changes are made. We also provide specifications of the ghost code functions readClock and incrementClock for manipulating the ghost resource for the global clock at ghost location γ_t. The function readClock requires fractional ownership of the clock resource for some non-zero fraction q to read the current clock value (line 4). On the other hand, incrementClock needs full ownership of the resource to increment the current clock value (line 5). Note that the functions lockNode and unlockNode follow the same specification as in the proof of search (see Figure 12.3).

With everything needed for the proof of upsert in place, let us now walk through the proof outline shown in Figure 12.5. We start with the invariant $\boxed{Inv(Inv_{LSM}, Prot)(r)}$, the view shift assumption on Prot, and the atomic precondition $\langle t\ H.\ MCS(r, t, H) \rangle$. The invariant can be accessed at each atomic step, but must also be reestablished after each step. Similarly, the atomic precondition is accessible at each atomic step, and must either be used to generate the postcondition of the atomic triple or the precondition must be reestablished.

1 $\{\mathsf{Node}(r, r, es, C_r)\}$

2   addContents $r\ k\ t$

3 $\{v.\ \mathsf{Node}(r, r, es, C_r')\ast C_r' = (v\ ?\ C_r[k \rightarrowtail t] : C_r)\}$

4 $\left\{\boxed{q\ \bullet\ t}^{\gamma_t}\ast q > 0\right\}$ readClock $()$ $\left\{v.\boxed{q\ \bullet\ t}^{\gamma_t}\ast v = t\right\}$

5 $\left\{\boxed{\bullet\ t}^{\gamma_t}\right\}$ incrementClock $()$ $\left\{\boxed{\bullet\ t+1}^{\gamma_t}\right\}$

6

7 $\left\{\boxed{\mathsf{Inv}(\mathsf{Inv}_{LSM}, \mathsf{Prot})(r)}\ast(\forall\ H\ k\ t.\ \mathsf{Prot}(H) \Rightarrow \mathsf{Prot}(H \cup \{(k,t)\}))\right\}\ast\left\langle t\ H.\ \mathsf{MCS}(r, t, H)\right\rangle$

8 **let** upsert $r\ k$ =

9   lockNode r;

10   $\{\mathsf{N_L}(r, r, C_r, Q_r)\}$

11   $\left\{\mathsf{Node}(r, r, es, C_r)\ast\cdots\ast\boxed{\circ\ C_r}^{\gamma_s}\ast\boxed{\tfrac{1}{2}C_r}^{\gamma_{c(r)}}\ast\boxed{\tfrac{1}{2}\ \bullet\ t_1}^{\gamma_t}\right\}$

12   **let** t = readClock $()$ **in**

13   $\left\{\begin{array}{l}\mathsf{Node}(r, r, es, C_r)\ast\cdots\ast\boxed{\circ\ C_r}^{\gamma_s}\ast\boxed{\tfrac{1}{2}C_r}^{\gamma_{c(r)}}\ast\boxed{\tfrac{1}{2}\ \bullet\ t_1}^{\gamma_t}\\\ast\ t = t_1\end{array}\right\}$

14   **let** res = addContents $r\ k\ t$ **in**

15   **if** res **then begin**

16     $\left\{\begin{array}{l}\mathsf{Node}(r, r, es, C_r')\ast\cdots\ast\boxed{\circ\ C_r}^{\gamma_s}\ast\boxed{\tfrac{1}{2}C_r}^{\gamma_{c(r)}}\ast\boxed{\tfrac{1}{2}\ \bullet\ t}^{\gamma_t}\\\ast\ C_r' = C_r[k \rightarrowtail t]\end{array}\right\}$

17   (* Linearization point *)

18     $\left\{\begin{array}{l}\mathsf{Node}(r, r, es, C_r')\ast\cdots\ast\boxed{\circ\ C_r}^{\gamma_s}\ast\boxed{\tfrac{1}{2}C_r}^{\gamma_{c(r)}}\ast\boxed{\tfrac{1}{2}\ \bullet\ t}^{\gamma_t}\\\ast\ C_r' = C_r[k \rightarrowtail t]\ast\mathsf{MCS}(r, t', H')\end{array}\right\}$

19     $\left\{\begin{array}{l}\mathsf{Node}(r, r, es, C_r')\ast\cdots\ast\boxed{\circ\ C_r}^{\gamma_s}\ast\boxed{\tfrac{1}{2}C_r}^{\gamma_{c(r)}}\ast\boxed{\tfrac{1}{2}\ \bullet\ t}^{\gamma_t}\\\ast\ C_r' = C_r[k \rightarrowtail t]\ast\mathsf{MCS}(r, t, H)\end{array}\right\}$

20   incrementClock $()$;

21     $\left\{\begin{array}{l}\mathsf{Node}(r, r, es, C_r')\ast\cdots\ast\boxed{\circ\ C_r}^{\gamma_s}\ast\boxed{\tfrac{1}{2}C_r'}^{\gamma_{c(r)}}\ast\boxed{\tfrac{1}{2}\ \bullet\ t+1}^{\gamma_t}\\\ast\ C_r' = C_r[k \rightarrowtail t]\ast H' = H \cup \{(k,t)\}\ast\mathsf{MCS}(r, t+1, H')\end{array}\right\}$

22     $\{\mathsf{N_L}(r, r, C_r', Q_r)\}\ast\left\langle\mathsf{MCS}(r, t+1, H \cup \{(k,t)\})\right\rangle$

23   unlockNode r

24     $\left\langle\mathsf{MCS}(r, t+1, H \cup \{(k,t)\})\right\rangle$

25   **end**

26   **else begin**

27     $\left\{\begin{array}{l}\mathsf{Node}(r, r, es, C_r')\ast\cdots\ast\boxed{\circ\ C_r}^{\gamma_s}\ast\boxed{\tfrac{1}{2}C_r}^{\gamma_{c(r)}}\\\ast\boxed{\tfrac{1}{2}\ \bullet\ t}^{\gamma_t}\ast C_r' = C_r\end{array}\right\}$

28     $\{\mathsf{N_L}(r, r, C_r, Q_r)\}$

29   unlockNode r;

30   upsert $r\ k$

31   **end**

32 $\left\langle\mathsf{MCS}(r, t+1, H \cup \{(k,t)\})\right\rangle$

Figure 12.5: Proof of upsert.

The thread first locks the root node, which transfers ownership of $N_L(r, r, es, C_r)$ from the invariant to the thread (line 10). At this point, we unfold the definition of $N_L(r, r, es, C_r)$ (line 11) as we will need the contained resources later in the proof. We here use the variable t_1 to refer to the current value of the global clock. Next, the thread uses `readClock` to read the current clock value into the local variable t. We can use the fractional permission $\boxed{\frac{1}{2} \bullet t_1}^{\gamma_t}$ and the specification of `readClock` to conclude $t = t_1$ (line 13). The thread now calls `addContents` to update r with the new copy of k for timestamp t. This leaves us with two possible scenarios depending on whether the return value *res* is *true* (line 16) or *false* (line 27).

In the case where `addContents` fails, no changes have been performed. Here, we simply fold $N_L(r, r, es, C_r)$ again, unlock r to transfer ownership of the node's resources back to the invariant, and commit the atomic triple on the recursive call to `upsert`.

In the case where `addContents` succeeds, we obtained $C_r' = C_r[k \rightarrowtail t]$ from its postcondition, where C_r' is the new contents of the root node. The thread will next call `incrementClock` to increase the global clock value. This will be the linearization point of this branch of the conditional expression. Hence, we will also have to update all remaining ghost resources to their new values at this point. To prepare committing the atomic triple, we first access its precondition to obtain $MCS(r, t', H')$ for some fresh variables t' and H' (line 18). We then open the invariant to access $MCS^\bullet(r, t, H)$ and use rule VIEW-SYNC to conclude $H' = H$ and $t' = t$ (line 19).

The actual commit of the atomic triple involves several steps. First, the thread calls `incrementClock`. To satisfy the precondition of `incrementClock`, we open the invariant to retrieve the second half of the fractional clock resource at ghost location γ_t, and combine it with the half in the thread's local state to obtain the full resource $\boxed{\bullet t}^{\gamma_t}$. The postcondition of `incrementClock` then gives us back $\boxed{\bullet t + 1}^{\gamma_t}$ which we split again into two halves, returning one half to the invariant and keeping the other in the local proof context. In addition, we update all remaining ghost resources:

- We update the authoritative version of the upsert history at ghost location γ_s in the invariant from H to $H' = H \cup \{(k, t)\}$, using rule AUTH-SET-UPD.

- We use rule FRAC-UPD to update the resource holding the root's contents at location $\gamma_{c(r)}$ from C_r to C_r' by reassembling the full resource from the half owned by the invariant, respectively, the half owned by the local proof context. After the update, the resource is split again into two halves, with one half returned to the invariant.

- We use the rule AUTH-MAXNAT-UPD to update the resource holding the root's contents-in-reach for k at ghost location $\gamma_{cir(r)(k)}$ from $B_r(k)$ to t. This is possible because $B_r = \bar{H}$ and $MaxTS(t, H)$ hold according to the invariant, which together imply $B_r(k) < t$.

- We use rule VIEW-UPD to update the client's and invariant's views of the data structures state to $MCS(r, t + 1, H')$ and $MCS^\bullet(r, t + 1, H')$, respectively.

This leaves us with the new proof context shown on line 21. It remains to show that the updates of the ghost resources preserve the invariant. That is, we need to prove that all constraints in the invariant involving t, H, C_r, and B_r' are maintained if we replace them with $t + 1$, H', C_r', and $B_r' = \lambda k.\, \bigl(C_r(k) \neq \bot\, ?\, C_r'(k) : Q_r(k)\bigr)$, respectively.

First note that we can use the view shift assumption on Prot to reestablish $\mathsf{Prot}(H')$ from $\mathsf{Prot}(H)$. Further note that by definition we have $B_r' = B_r[k \mapsto t]$. Hence, the update to the ghost location $cir(r)(k)$ preserved the invariant. We can similarly see that the updates to γ_t, γ_S, and γ_{cr} preserve the invariant.

Next, observe that $\mathsf{MaxTS}(t + 1, H')$ follows directly from the definition of H' and $\mathsf{MaxTS}(t, H)$. We obtain the later from the invariant prior to the call to `incrementClock`.

To show $\bar{H}' = B_r'$, we need to prove that for all keys k'

$$\bar{H}'(k') = \bigl(C_r'(k') \neq \bot\, ?\, C_r'(k') : Q_r(k')\bigr)$$

If $k' \neq k$, the equality follows directly from $\bar{H} = B_r$ and the definitions of H' and C_r'. For the case where $k' = k$ observe that $\mathsf{MaxTS}(t, H)$ implies $\bar{H}'(k) = t$. Moreover, we have $C_r'(k) = t$ by definition of C_r' and we also know $t \neq \bot$ because the clock resource $\gamma(t)$ can hold only natural numbers.

Finally, observe that $\phi_2(r, C_r', Q_r, I_r)$ holds trivially because $I_r.in = \lambda_0$. Hence, we conclude that the updates maintain the invariant.

We can continue with the remainder of the proof. To prove that we can safely unlock r, we have to reassemble $\mathsf{N_L}(r, r, C_r')$ from the proof context at line 21. We have all the relevant pieces available, except for $\overline{\circ\, C_r'}^{\gamma_S}$. We obtain this remaining piece by observing that $\overline{\circ\, C_r}^{\gamma_S}$ implies $C_r \subseteq H'$ which in turn implies $C_r' \subseteq H'$ by definition of C_r' and H'. Using rule AUTH-SET-SNAP we obtain $\overline{\circ\, H'}^{\gamma_S}$, which we can rewrite to $\overline{\circ\, (H' \cup C_r')}^{\gamma_S}$ using the previously derived equality $H' = H' \cup C_r'$. Applying rule AUTH-FRAG-OP, we can then infer $\overline{\circ\, H'}^{\gamma_S} * \overline{\circ\, Cr'}^{\gamma_S}$ and after throwing away the first conjunct, we are left with the desired missing piece.

After reassembling $\mathsf{N_L}(r, r, es, C_r')$ we arrive at line 22 at which point the thread unlocks r, transferring $\mathsf{N_L}(r, r, es, C_r')$ back to the invariant. This concludes the proof of `upsert`.

Our careful encoding of contents-in-reach ensured that the sets Q_n and the node-local interfaces I_n are not affected by the upsert for any node n, including r. This considerably simplified the proof that the invariant $\mathsf{Inv}(\mathsf{Inv}_{LSM}, \mathsf{Prot})(r)$ is maintained at the linearization point.

12.3 MULTICOPY MAINTENANCE OPERATIONS

The true complexity in a correctness proof of a concurrent data structure often arises only when maintenance operations are taken into account (e.g., `flush` or `compact` in the case of the LSM tree). We next show that we can extend our multicopy structure template with a generic maintenance operation without substantially increasing the proof complexity. The basic idea of our proofs here is that for every timestamped copy of key k, denoted as the pair (k, t), every maintenance operation either does not change the distance of (k, t) to the root or increases it while preserving an edgeset-guided path to (k, t). Using these two facts, we can prove that all the structure invariants are also preserved.

12.3.1   MAINTENANCE TEMPLATE

For the maintenance template, we consider a generalization of the compaction operation found in LSM tree implementations such as LevelDB [Google, 2020] and Apache Cassandra [Apache Software Foundation, 2020a, Jonathan Ellis, 2011]. While those implementations work on lists for the high-level multicopy structure, our maintenance template supports arbitrary tree-like multicopy structures. The code is shown in Figure 12.6. The template uses the helper function atCapacity n to test whether the size of n (i.e., the number of non-$\bot$ entries in n's contents) exceeds an implementation-specific threshold. If not, then the operation simply terminates. In case n is at capacity, the function chooseNext is used to determine the node to which the contents of n can be merged. If the contents of n can be merged to successor m of n, then chooseNext returns Some m. In case no such successor exists, then it returns None. If chooseNext returns Some m, then the contents of n are merged to m. By merge, we mean that some copies of keys are transferred from n to m, possibly replacing older copies in m. The merge is performed by the helper function mergeContents. It must ensure that all keys k merged from C_n to C_m satisfy $k \in \text{es}(n, m)$.

On the other hand, if chooseNext returns None, then a new node is allocated using the function allocNode. Once the new node is allocated, it is added to the data structure using the helper function insertNode. Here, the new edgeset $\text{es}(n, m)$ must be disjoint from all edgesets for the other successors m' of n. Afterward, the contents of n are merged to m as before. Note that the maintenance template never removes nodes from the structure. In practice, the depth of the structure is bounded by letting the capacity of nodes grow exponentially with the depth. The right-hand side of Figure 12.6 shows the intermediate states of a potential execution of the compact operation.

12.3.2   PROOF OF MAINTENANCE TEMPLATE

High-Level Proof Idea.   To verify compact we need to prove that the following proposition is valid:

$$\boxed{\text{Inv}(\text{Inv}_{LSM}, \text{Prot})} \twoheadrightarrow \big\langle t\ H.\ \text{MCS}(r, t, H) \big\rangle \text{ compact } r \ \big\langle \text{MCS}(r, t, H) \big\rangle$$

This proposition says that compact logically takes effect in a single atomic step, and at this step the abstract state of the data structure does not change and its invariant is preserved.

Technically, the linearization point of the operation occurs when all locks are released, just before the function terminates. However, the interesting part of the proof is to show that the changes to the physical contents of nodes n and m performed by each call to mergeContents at line 7 preserve the abstract state of the structure as well as the invariants. In particular, the

```
1  let rec compact n =
2    lockNode n;
3    if atCapacity n then begin
4      match chooseNext n with
5      | Some m ->
6        lockNode m;
7        mergeContents n m;
8        unlockNode n;
9        unlockNode m;
10       compact m
11     | None ->
12       let m = allocNode () in
13       insertNode r n m;
14       mergeContents n m;
15       unlockNode n;
16       unlockNode m;
17       compact m
18   end
19   else
20     unlock n
```

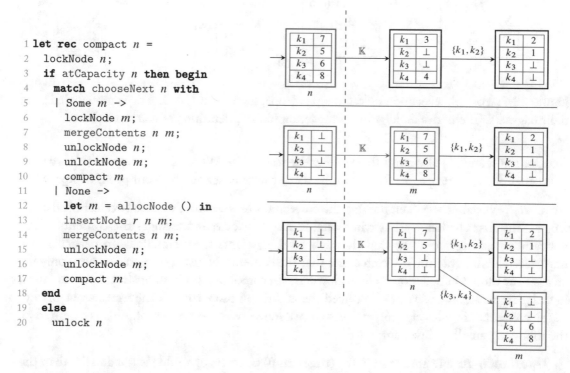

Figure 12.6: Maintenance template for tree-like multicopy structures. The template can be instantiated by providing implementations of helper functions atCapacity, chooseNext, mergeContents, allocNode, and insertNode. atCapacity n returns a Boolean value indicating whether node n has reached its capacity. The helper function chooseNext n returns Some m if there exists a successor m of n in the data structure into which n should be compacted, and None in case n cannot be compacted into any of its successors. mergeContents n m (partially) merges the contents of n into m. Finally, allocNode is used to allocate a new node and insertNode n m inserts node m into the data structure as a successor of n. The right-hand side shows a possible execution of compact. Edges are labeled with their edgesets. The nodes n and m in each iteration are marked in blue.

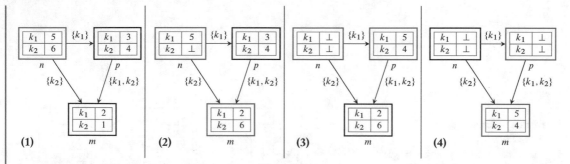

Figure 12.7: Possible execution of the `compact` operation on a DAG. Edges are labeled with their edgesets. The nodes undergoing compaction in each iteration are marked in blue.

changes to C_n and C_m also affect the contents-in-reach of m. We need to argue that this is a local effect that does not propagate further in the data structure, as we did in our proof of `upsert`.

Auxiliary invariants. When proving the correctness of `compact`, we face two technical challenges. The first challenge arises when establishing that `compact` changes the contents of the nodes involved in such a way that the high-level invariants are maintained. In particular, we must reestablish Invariant 2, which states that the contents-in-reach of each node can only increase over time. Compaction replaces downstream copies of keys with upstream copies. Thus, in order to maintain Invariant 2, we need the additional property that the timestamps of keys in the contents of nodes can only decrease as we move away from the root. This is captured by the following auxiliary invariant:

Invariant 6 At any atomic step, the (timestamp) contents of a node is not smaller than the contents-in-reach of its successor. That is, for all keys k and nodes n and m, if $k \in \text{es}(n,m)$ and $C_n(k) \neq \perp$ then $C_{ir}(m)(k) \leqslant C_n(k)$.

We can capture Invariant 6 in our data structure invariant $\text{MCS}(r, t, H)$ by adding the following predicate as an additional conjunct to the predicate $\text{N}_\text{S}(r, n, C_n, Q_n, H)$:

$$\phi_3(n) := \forall k.\, Q_n(k) \leqslant B_n(k) \tag{12.6}$$

The second challenge is that the maintenance template generates only tree-like structures. This implies that at any time there is at most one path from the root to each node in the structure. We will see that this invariant is critical for maintaining Invariant 6. However, the data structure invariant presented thus far allows for arbitrary DAGs.

To motivate this issue further, consider the multicopy structure in step **(1)** of Figure 12.7. The logical contents of this structure (i.e., the contents-in-reach of n) is $\{k_1 \rightarrowtail 5, k_2 \rightarrowtail 6\}$.

The structure in step **(2)** shows the result obtained after executing `compact` n to completion where n has been considered to be at capacity and the successor m has been chosen for the merge, resulting in $(k_2, 6)$ being moved from n to m. Note that at this point the logical contents of the data structure is still $\{k_1 \rightarrowtail 5, k_2 \rightarrowtail 6\}$ as in the original structure. However, the structure now violates Invariant 6 for nodes p and m since $B_m(k_2) > C_p(k_2)$.

Suppose that now a new compaction starts at n that still considers n at capacity and chooses p for the merge. The merge then moves the copy$(k_1, 5)$ from n to p. The graph in step **(3)** depicts the resulting structure. The compaction then continues with p, which is also determined to be at capacity. Node m is chosen for the merge, resulting in $(k_1, 5)$ and $(k_2, 4)$ being moved from p to m. At this point, the second compaction terminates. The final graph in step **(4)** shows the structure obtained at this point. Observe that the logical contents is now $\{k_1 \rightarrowtail 5, k_2 \rightarrowtail 4\}$. Thus, this execution violates the specification of `compact`, which implies that the logical contents must be preserved. In fact, the contents-in-reach of n has decreased, which violates Invariant 2.

The example illustrates that DAGs appear to be problematic for the correctness of the maintenance template. If we strengthened the data structure invariant to state that the graph is a tree, then this would allow us to prove that Invariant 6 is maintained, which in turn guarantees that Invariant 2 is maintained when copies of keys are merged downstream.

We observe that although `compact` will create only tree-like structures, we can prove its correctness using a weaker invariant that does not rule out non-tree DAGs, but instead focuses on how `compact` interferes with concurrent `search` operations. This weaker invariant relies on the fact that for every key k in the contents of a node n, there exists a unique search path from the root r to n for k. That is, if we project the graph to only those nodes reachable from the root via edges (n, m) that satisfy $k \in \text{es}(n, m)$, then this projected graph is a list. Using this weaker invariant we can capture implementations based on B-link trees or skip lists which are DAGs but have unique search paths.

To this end, recall from §7.4 the notion of the *inset* of a node n, $\text{ins}(n)$, which is the set of keys k such that there exists a (possibly empty) path from the root r to n, and k is in the edgeset of all edges along that path. That is, since a `search` for a key k traverses only those edges (n, m) in the graph that have k in their edgeset, the `search` traverses (and accesses the contents of) only those nodes n such that $k \in \text{ins}(n)$. Now observe that `compact`, in turn, moves new copies of a key k downward in the graph only along edges that have k in their edgeset. The following invariant is a consequence of these observations and the definition of contents-in-reach:

Invariant 7 At any atomic step, a key is in the contents-in-reach of a node only if it is also in the node's inset. That is, for all keys k and nodes n, if $C_{ir}(n)(k) \neq \bot$ then $k \in \text{ins}(n)$.

This invariant rules out the problematic structure in step **(1)** of Figure 12.7 because we have $C_{ir}(p)(k_2) = 4$ but $k_2 \notin \text{ins}(p) = \{k_1\}$.

Invariant 7 alone is not enough to ensure that Invariant 6 is preserved. For example, consider the structure obtained from **(1)** of Figure 12.7 by changing the edgeset of the edge (n, p) to $\{k_1, k_2\}$. This modified structure satisfies Invariant 7 but allows the same problematic execution ending in the violation of Invariant 6 that we outlined earlier. However, observe that in the modified structure $k_2 \in \mathsf{es}(n, p) \cap \mathsf{es}(n, m)$, which violates the property that all edgesets leaving a node are disjoint. We have already captured this property in our data structure invariant (as an assumption on the implementation-specific predicate $\mathsf{Node}(r, n, es, C_n)$). However, in our formal proof we need to rule out the possibility that a search for k can reach a node m via two *incoming* edgesets $\mathsf{es}(n, m)$ and $\mathsf{es}(p, m)$. Proving that disjoint *outgoing* edgesets imply unique search paths involves global inductive reasoning about the paths in the multicopy structure. To do this using only local reasoning, we will instead rely on an inductive consequence of locally disjoint outgoing edgesets, which we capture explicitly as an additional auxiliary invariant (and which we will enforce using flows):

> **Invariant 8** At any atomic step, the distinct immediate predecessors of any node n have disjoint insets. More precisely, for all distinct nodes n, p, m, and keys k, if $k \in \mathsf{es}(n, m) \cap \mathsf{es}(p, m)$ then $k \notin \mathsf{ins}(n) \cap \mathsf{ins}(p)$.

Note that changing the edgeset of (n, p) in Figure 12.7 to $\{k_1, k_2\}$ would violate Invariant 8 because the resulting structure would satisfy $k_2 \in \mathsf{es}(n, m) \cap \mathsf{es}(p, m)$ and $k_2 \in \mathsf{ins}(n) \cap \mathsf{ins}(p)$.

In order to capture invariants 7 and 8 in $\mathsf{MCS}(r, t, H)$, we introduce an additional flow that we use to encode the inset of each node. The encoding of insets in terms of a flow follows §7.4. That is, the underlying flow domain is multisets of keys $M = \mathbb{K} \to \mathbb{N}$ and the actual calculation of the insets is captured by (FlowEqn) if we define:

$$e(n, n') := \lambda m.\, m \cap \mathsf{es}(n, n') \qquad\qquad in(n) := \chi\,(n = r \,?\, \mathbb{K} : \emptyset)$$

If $\mathit{fl}_{\mathsf{ins}}$ is a flow that satisfies (FlowEqn) for these definitions of e and in, then for any node n that is reachable from r, $\mathit{fl}_{\mathsf{ins}}(n)(k) > 0$ iff $k \in \mathsf{ins}(n)$. Invariants 7 and 8 are then captured by the following two predicates, which we add to $\mathsf{N_S}$:

$$\phi_4(n) := \forall k.\, B_n(k) = \bot \vee \mathit{fl}_{\mathsf{ins}}(n)(k) > 0 \qquad\qquad \phi_5(n) := \forall k.\, \mathit{fl}_{\mathsf{ins}}(n)(k) \leqslant 1$$

Note that ϕ_5 captures Invariant 8 as a property of each individual node n by taking advantage of the fact that the multiset $\mathit{fl}_{\mathsf{ins}}(n)$ explicitly represents all of the contributions made to the inset of n by n's predecessor nodes.

We briefly explain why we can still prove the correctness of search and upsert with the updated data structure invariant. First note that search does not modify the contents, edgesets, or any other ghost resources of any node. So the additional conjuncts in the invariant are trivially maintained.

Now let us consider the operation upsert $r\,k$. Since upsert does not change the edgesets of any nodes, the resources and constraints related to the inset flow are trivially maintained, with the exception of $\phi_4(r)$: after the upsert we have $B_r(k) \neq \bot$ which may not have been true before. However, from $in(r)(k) = 1$, the flow equation, and the fact that the flow domain is positive, it follows that we must also have $\mathit{fl}_{\mathsf{ins}}(r)(k) > 0$ (i.e., $k \in \mathsf{ins}(r) = \mathbb{K}$). Hence, $\phi_4(r)$ is preserved as well.

Proof outline of `compact`.   As discussed previously, we need to extend the data structure invariant Inv_{LSM} with ghost resources that track the inset of each node. We do this via an additional flow interface. The new global interface, denoted J, is stored at a new ghost location γ_J in Inv_{LSM}. As for the contents-in-reach flow interface I, the associated RA is authoritative flow interfaces over the flow domain of multisets of key-timestamp pairs. We add the constraint $\mathsf{dom}(I) = \mathsf{dom}(J)$ to ensure that I and J agree on which nodes belong to the graph.

The actual calculation of the insets is captured by the following constraint on the outflow of the singleton interfaces J_n which we add to the predicate $\mathsf{N_S}$:

$$J_n.out = \lambda n'\, k.\ \big(k \in es_n(n')\ ?\ J_n.in(n)(k) : 0\big) \tag{12.7}$$

Additionally, we add the following constraint to G, which requires that the global interface J gives the full keyspace as inflow to the root node r and no inflow to any other node:

$$J.in = \lambda n\, k.\ (n = r\ ?\ 1 : 0)$$

Together, these constraints guarantee that for any node n that is reachable from r, $J_n.in(k) > 0$ iff $k \in \mathsf{ins}(n)$.

Finally, we add the following predicates to $\mathsf{N_S}$ in order to capture invariants 6-8:

$$\phi_3(n, C_n, Q_n) := \forall k.\ Q_n(k) \leqslant B_n(k)$$
$$\phi_4(n, C_n, Q_n, J_n) := \forall k.\ B_n(k) = \bot \vee J_n.in(n)(k) > 0$$
$$\phi_5(n, J_n) := \forall k.\ J_n.in(n)(k) \leqslant 1$$

The specifications of the implementation-specific helper functions assumed by `compact` are provided in Figure 12.8. A thread performing `compact n` starts by locking node n and checking if node n is at full capacity using the helper function `atCapacity`. By locking node n, the thread receives the resources available in $\mathsf{N_L}(r, n, C_n, Q_n)$, for some contents C_n and successor contents-in-reach Q_n. The precondition of `atCapacity` requires the predicate $\mathsf{Node}(r, n, es_n, C_n)$, which is available to the thread as part of $\mathsf{N_L}(r, n, C_n, Q_n)$. The return value of `atCapacity n` is a Boolean indicating whether node n is full or not. The precise logic of how the implementation of `atCapacity` determines whether a node is full is immaterial to the correctness of the template, so the specification of `atCapacity` abstracts from this logic. If n is not full, `compact` releases the lock on n, transferring ownership of $\mathsf{N_L}(r, n, C_n, Q_n)$ back to the invariant and then terminates. The call to `unlockNode` on line 20 is the commit point of the atomic triple in the else branch of the conditional.

Thus, let us consider the other case, i.e., when n is full. Here, the contents of node n must be merged with the contents of some successor node m of n. This node m is determined by the helper function `chooseNext`. `chooseNext` either returns an existing successor m of n (i.e., $es'_n(m) \neq \emptyset$), or n needs a new successor node, which we capture by the implementation-specific predicate `needsNewNode`(r, n, es, C_n). In the former case, we can establish that m is part of the data structure due to the fact that the edgeset of n directs some keys to m. This

1 $\{\mathsf{Node}(r, n, es, C_n)\}$
2 `atCapacity` n
3 $\{b.\, \mathsf{Node}(r, n, es, C_n) * b = \mathit{true} \vee b = \mathit{false}\}$
4
5 $\{\mathsf{Node}(r, n, es_n, C_n)\}$
6 `chooseNext` n
7 $\left\{\begin{array}{c} v.\, \mathsf{Node}(r, n, es_n, C_n) * (v = \mathsf{Some}(m) * es_n'(m) \neq \emptyset \\ \vee\, v = \mathsf{None} * \texttt{needsNewNode}(r, n, es, C_n)) \end{array}\right\}$
8
9 $\{\mathsf{True}\}$
10 `allocNode ()`
11 $\{m.\, \mathsf{Node}(r, m, \lambda_\perp, \lambda_\emptyset)\}$
12
13 $\{\mathsf{Node}(r, n, es_n, C_n) * \texttt{needsNewNode}(r, n, es_n, C_n) * \mathsf{Node}(r, m, \lambda_\perp, \lambda_\emptyset)\}$
14 `insertNode` r n m
15 $\{\mathsf{Node}(r, n, es_n', C_n) * \mathsf{Node}(r, m, \lambda_\perp, \lambda_\emptyset) * es_n' = es_n[m \rightarrowtail es_n'(m)] * es_n'(m) \neq \emptyset\}$
16
17 $\{\mathsf{Node}(r, n, es_n, C_n) * \mathsf{Node}(r, m, es_m, C_m) * es_n(m) \neq \emptyset\}$
18 `mergeContents` n m
19 $\left\{\begin{array}{c} \mathsf{Node}(r, n, es_n, C_n') * \mathsf{Node}(r, m, es_m, C_m') \\ * C_n' \subseteq C_n * C_m' \subseteq C_n \cup C_m * C_n \cap C_m' \cap C_n' = \emptyset * \mathrm{dom}(C_m) \subseteq \mathrm{dom}(C_m') \\ * \mathit{merge}(C_n, es_n(m), C_m) = \mathit{merge}(C_n', es_n(m), C_m') \end{array}\right\}$

Figure 12.8: Specifications of helper functions used by `compact`, `atCapacity` n, `chooseNext` n, and `mergeContents` $n\, m$.

follows from the property $\texttt{closed}(n)$ in the invariant $\mathsf{Inv}(\mathsf{Inv}_{LSM}, \mathsf{Prot})(r)$. In the latter case, a new node is allocated using the helper function `allocNode` and inserted into the data structure as a successor of n using the helper function `insertNode`. After this call, m becomes reachable from the root node r from n. To ensure that the node-local invariant of n is maintained, m must be "registered" in $\mathsf{Inv}(\mathsf{Inv}_{LSM}, \mathsf{Prot})(r)$. To this end, we must extend the domain of the global flow interfaces tracked by Inv_{LSM} with the new node m. This can be done using a frame-preserving update of the authoritative version of the interfaces at ghost locations γ_I and γ_J. Showing that the invariant is preserved after this update is easy because the postcondition of `insertNode` together with the invariant guarantee that m has no outgoing edges and can only be reached via n. In particular, the fact that $C_m = \lambda_\perp$ and $es_m = \lambda_\emptyset$ imply in this case that $\phi_4(m, C_m, Q_m, J_m)$ holds, where $B_m = C_m$. Additionally, the conjunct $es_n' = es_n[m \rightarrowtail es_n'(m)]$ ensures that the inset of any nodes other than n and m do not change. Hence, the flow interfaces at γ_I and γ_J can be contextually extended to include node m in their domains.

Overall, once it has been established that node m is in the domain of the global flow interfaces I and J in the invariant $\mathsf{Inv}(\mathsf{Inv}_{LSM}, \mathsf{Prot})(r)$, m is locked and the contents of n is (partially) merged into m using the helper function `mergeContents` (line 7).

Let us now examine the specification of mergeContent in detail. mergeContents $n\, m$ merges the data from node n into node m. By merge, we mean that some copies of keys are transferred from n to m, possibly replacing older copies in m. In general, mergeContents modifies the contents of the two nodes according to the specification given in Figure 12.8. The precondition demands ownership of the physical representation of the nodes' contents and that m is a successor of n. The contents are modified to C'_n and C'_m, respectively, such that

- no new copies are added n: $C'_n \subseteq C_n$,

- only copies from n are added to m: $C'_m \subseteq C_n \cup C_m$,

- copies moved from n to m are not left in n: $C_n \cap C'_m \cap C'_n = \emptyset$, and

- no keys are removed from m: $\mathsf{dom}(C_m) \subseteq \mathsf{dom}(C'_m)$.

To further constrain the new contents of the nodes, the postcondition additionally demands $merge(C_n, es_n(m), C_m) = merge(C'_n, es_n(m), C'_m)$, where

$$merge(C_n, Es, C_m) := \lambda k.\ \begin{cases} C_n(k) & \text{if } C_n(k) \neq \bot \\ C_m(k) & \text{else if } k \in Es \\ \bot & \text{otherwise} \end{cases}$$

We next explain that, together, these constraints ensure that we can consistently update all relevant ghost resources in the invariant. In particular, the contents-in-reach of m can only increase and the contents-in-reach of all other nodes, including n remain unchanged.

For any key k and contents C, we denote by $\mathsf{dom}(C)$ the set of keys k such that $C(k) \neq \bot$. The set $K := \mathsf{dom}(C_n) \setminus \mathsf{dom}(C'_n)$ then denotes all keys whose copies are merged from C_n into C_m.

Observe that the last conjunct in the postcondition of mergeContents guarantees that only copies of keys in the edgeset of the edge (n, m) are merged. That is, if $k \in K$, then $k \in es_n(m)$. This holds because, if $k \in \mathsf{dom}(C_n) \setminus \mathsf{dom}(C'_n)$ and $k \notin es_n(m)$, then $merge(C'_n, es_n(m), C'_m)(k) = \bot$, but $merge(C_n, es_n(m), C_m)(k) = C_n(k) \neq \bot$, which is a contradiction. We will use this observation freely in the remainder of the proof.

Before we proceed with the rest of the proof, we fix instantiations for the existentially quantified variables in the invariant $\mathsf{Inv}(\mathsf{Inv}_{LSM}, \mathsf{Prot})(r)$ when we assume it before the call to mergeContents. For a node p in the structure, we denote by I_p and J_p the fragmental singleton flow interfaces of node p at ghost locations γ_I and γ_J, respectively. Moreover, let Q_p be the set stored at ghost location $\gamma_{q(p)}$ (i.e., the successor contents-in-reach of p before the call to mergeContents).

First, since the update of C_m also affects the contents-in-reach of m, we need to update the resource storing B_m appropriately. We do this by defining:

$$B'_m := \lambda k.\ (k \in K\ ?\ C_n(k) : B_m)$$

and then replace for each key k, $B_m(k)$ at ghost location $\gamma_{C_{ir}(m)(k)}$ by $B'_m(k)$. To do this, the authoritative maxnat RA requires us to show that $B_m(k) \leqslant B'_m(k)$. If $k \notin K$, then $B_m(k) = B'_m(k)$ by definition. Hence consider the case where $k \in K$. We then have $B'_m(k) = C_n(k) \neq \bot$. From this and $k \in e_n(m)$, it follows that $I_n.out(m)(k, Q_n(k)) > 0$, which, using the flow equation, enables us to conclude $J_n.in(m)(k, Q_n(k)) > 0$. We then infer from $\phi_2(m, C_m, Q_m, I_m)$ that $Q_n(k) = B_m(k)$. Moreover, it follows from $\phi_3(n, C_n, Q_n)$ that $Q_n(k) \leqslant B_n(k)$. Since $B_n(k) = C_n(k) = B'_m(k)$, we can conclude that $B_m(k) \leqslant B'_m(k)$ as desired.

Second, we observe that since $\texttt{mergeContents}\, n\, m$ does not change the edgesets of any nodes, the insets of all nodes also remain unchanged. In particular, for all nodes p, the singleton interfaces J_p are unaffected and hence still compose to the global interface R. Additionally, this means that $\phi_5(p, J_p)$ and the constraint (12.7) on $J_p.out$ are trivially maintained. To see that $\phi_4(m, C'_m, Q_m, J_m)$ is also preserved, we note that if $B_m(k) = \bot$ and $B'_m(k) \neq \bot$ for some k, then $k \in K$ and hence $C_n(k) = B_n(k) = B'_m(k)$. It then follows from $\phi_4(n, C_n, Q_n, J_n)$ that $J_n.in(n)(k) > 0$. Moreover, $k \in K$ implies $k \in es_n(m)$ and, hence, $J_n.out(m)(k) > 0$. It then follows from the flow equation that $J_m.in(m)(k) > 0$.

Finally, we need to reflect the change in the contents-in-reach of m in the local invariant of n by updating the ghost resource holding Q_n to the new value

$$Q'_n = \lambda k.\ (k \in K ?\, C_n(k) : Q_n(k))$$

In turn, this requires an update of the singleton interfaces I_n and I_m to I'_n and I'_m such that:

$$
\begin{aligned}
I'_n.in &:= I_n.in \\
I'_n.out &:= \lambda n'\,(k, t).\ I_n.out(n')(k, t) \\
&\qquad\qquad + (n' = m \wedge k \in K \wedge C_n(k) = t ?\, 1 : 0) \\
I'_m.in &:= \lambda n'\,(k, t).\ I_m.in(n')(k, t) \\
&\qquad\qquad + (n' = m \wedge k \in K \wedge C_n(k) = t ?\, 1 : 0) \\
I'_m.out &:= I_m.out
\end{aligned}
$$

First, note that the changes to the inflows and outflows match up consistently. One can therefore easily verify that the old and new singleton interfaces compose to the same larger two-node interface:

$$I_n \oplus I_m = I'_n \oplus I'_m$$

This means that we can simultaneously replace all old interfaces by their new ones using a frame-preserving update of ghost location γ_I.

For the complete proof of the maintenance operation, we defer the interested reader to our Iris development (cf. Chapter 13).

12.4   VERIFYING IMPLEMENTATIONS

Similar to the single-copy template algorithms we have considered in Chapter 8, to obtain a verified implementation of the LSM DAG template, one only needs to specify the concrete representation of a node (i.e., define the predicate Node) and provide implementations of the helper functions that satisfy the specifications assumed by the proofs of the template algorithms (e.g., the functions `findNext` and `inContents` for `search`). We note that the template proofs can accommodate implementations that polymorphically dispatch a helper function call to different implementations of the function based on whether the call operates on the root node, which typically resides in memory, or a non-root node, which typically resides on disk. We discuss one such implementation briefly in the next chapter.

CHAPTER 13

Proof Mechanization and Automation

This chapter evaluates the techniques presented so far by mechanically verifying the template algorithms from Chapters 8 - 12, as well as some real-world implementations based on them.

For the single-copy structures, we verify the give-up, link, and lock-coupling template proofs that were described in Chapter 8. For the multicopy structures, we verify the LSM DAG template from Chapter 12 and a two-node multicopy template for differential files described in Chapter 9. Additionally for multicopy structures, we have also mechanized the reasoning about the client-level specification assuming the template-level specification (as per the discussion in Chapter 11). Note that this reasoning makes no assumptions about the implementations of the multicopy structure operations, other than that they satisfy the template-level specifications. Hence, the proof that the template-level specifications imply the client-level specifications can be reused across different templates for multicopy structures, such as the two specific templates that we verify here.

These proofs have been mechanically checked using the Coq proof assistant, building on the formalization of Iris [Jung et al., 2015, Krebbers et al., 2018, 2017b]. The proofs parameterize over the implementation of the helper functions (e.g., decisiveOp, findNext, etc.) and the heap representation predicate Node.

For the give-up and link templates, we have derived and verified implementations based on B-trees and hash tables. In particular, we verify the B-link tree implementation described in Chapter 6. For the lock-coupling template, we have considered a sorted linked list implementation. For the multicopy structure templates, we verify an LSM tree implementation to demonstrate an application of the LSM DAG template. The work to verify an implementation of an LSM B-link tree (a generalization of the LSM tree that utilizes the DAG structures permitted by the template) and an implementation of a differential file for the two-node template is ongoing. Our case studies are summarized in Figure 13.1.

These concrete implementations have been verified using the separation logic based deductive program verifier GRASShopper [Piskac et al., 2014]. As the tool uses SMT solvers to largely automate the verification process, this provided us with a substantial decrease in effort. We do not have, as of now, a formal proof for the transfer of proofs between Iris and GRASShopper. Iris support all the reasoning that we do in GRASShopper. Thus, the imple-

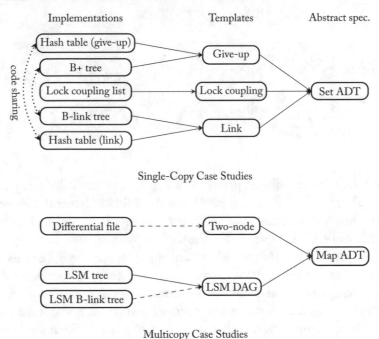

Figure 13.1: The templates and implementations in our case studies. The arrows indicate that all the single-copy templates satisfy the abstract specification of a Set ADT (as defined in Figure 8.1), while the multicopy templates satisfy the abstract Map ADT. The arrows also connect the implementations to the template that they instantiate. The dashed lines represent ongoing work.

mentation proofs can also be carried out in Iris to obtain end-to-end verification. However, this comes with significant additional manual effort.

With respect to single-copy structures, we have also verified the split operations for B-link trees. The B-link tree uses a two-part split operation: a half-split that creates a new node, transfers half the contents from a full node to this new node, then adds a link edge; and a full-split that completes the split by linking the original node's parent to the new node. For the split operations, we assume a harness template for a maintenance thread that traverses the data structure graph to identify nodes that are amenable to half splits. While we have not verified this harness, we note that it is a variation of our lock-coupling template where the abstract specification leaves the contents of the data structure unchanged. For the implementations of half and full splits, we verify that the operation preserves the flow interface of the modified region as well as its contents.

The LSM List implementation that we verify uses an unsorted array to store key-timestamp pairs for the root node (with upserts appended to one end of the array), and a read-only sorted array (also known as a sorted string table [Google, 2020]) for the other (on-disk) nodes. For this implementation, we verify the helper functions needed by the core search structure operations (Figure 12.1), but not those needed by the maintenance template (Figure 12.6). We do not anticipate any difficulty in extending the implementation proof to also support compaction.

Our verification effort includes a mechanization of the meta-theory of flows including that flow interfaces form an RA (Theorem 7.9) and allow frame-preserving updates according to rule FLOWINT-DOM-UPD. Our formalization is parametric in the flow domain (i.e., the underlying cancellative, commutative monoid). We also provide instantiation of the meta-theory for the specific flow domains used in our proofs (e.g., multisets). We have duplicated this effort in Iris/Coq and GRASShopper in order to make the two parts of our verification self-contained. The formalization is available as two standalone libraries that can be reused for other flow-based proofs in these systems. The full development of our mechanization effort is available online[1].

Table 13.1 provides a summary of our development. Experiments have been conducted on a laptop with an Intel Core i7-8750H CPU and 16GB RAM. We split the table into one part for the templates (proved in Coq) and one part for the implementations (proved in GRASShopper). We note that for the B-link tree, B+ tree, hash table implementations and LSM implementation, most of the work is done by the array library, which is shared between all these data structures. The size of the proof for the lock-coupling list and maintenance operations is relatively large. The reason is that these involve the calculation of a new flow interface for the region obtained after the modification. This requires the expansion of the definitions of functions related to flow interfaces, which are deeply nested quantified formulas. GRASShopper enforces strict rules that limit quantifier instantiation so as to remain within certain decidable logics [Bansal et al., 2015, Piskac et al., 2013]. Most of the proof in this case involves auxiliary assertions that manually unfold definitions. The size of the proof could be significantly reduced with a few improved tactics for quantifier expansion [Leino and Pit-Claudel, 2016].

Our proof effort began with the goal of verifying the link template algorithm. It is difficult to assess the overall time effort spent on verifying the link template algorithm because we designed our verification methodology as we verified the template. However, with all the machinery now in place, our experience is that verifying a new template algorithm is a matter of a few hours of proof effort. In fact, adapting the link template proof to the give-up template was straightforward and required only minor changes. Our experience with adapting implementation proofs is similar.

The template algorithms that we have verified focus on lock-based techniques for thread synchronization. In fact, many real-world applications perform better using lock-based algorithms instead of lock-free algorithms as the latter tend to copy data more[2]. On the other hand,

---

[1]https://github.com/nyu-acsys/template-proofs/tree/css_book

[2]For instance, Apache's CouchDB uses a B+ tree with a global write lock; BerkeleyDB, which has hosted Google's account information, uses a B+ tree with page-level locks in order to trade-off concurrency for better recovery; and java.util.concurrent's hash tables lock the entire list in a bucket during writes, which is more coarse-grained than the one we verify.

Table 13.1: Summary of templates and instantiations verified in Iris/Coq and GRASShopper. For each algorithm or library, we show the number of lines of code, lines of proof annotation (including specification), total number of lines, and the proof-checking/verification time in seconds.

Module	Code	Proof	Total	Time
Templates (Iris/Coq)				
Flows Library	0	3757	3757	42
Lock Implementation	10	103	113	5
Single-copy:				
Link Template	21	638	659	80
Give-up Template	26	453	479	28
Lock-coupling Template	34	1057	1091	111
Multicopy:				
Client-level Spec	2	1022	1024	36
LSM DAG Template	46	3586	3632	312
Two-node Template	26	942	968	65
Total	**165**	**11558**	**11723**	**674**
Implementations (GRASShopper)				
Flows Library	5	717	722	6
Array Library	143	322	465	6
Single-copy:				
B+ tree	63	99	162	11
B-link (core)	85	161	246	23
B-link (half split)	34	192	226	75
B-link (full split)	17	137	154	594
Hash table (link)	54	99	153	8
Hash table (give-up)	60	138	198	10
Lock-coupling List	59	300	359	37
Multicopy:				
LSM Tree	110	221	331	5
Total	**630**	**2386**	**3016**	**831**

our methodology does not require locking, and can be extended to prove data structures with lock-free algorithms such as the Bw-tree [Levandoski and Sengupta, 2013]. We discuss such extensions as part of the future work in §14.2.

CHAPTER 14

Related Work, Future Work, and Conclusion

14.1 RELATED WORK

Correctness criteria for concurrent data structures. A basic correctness criterion for concurrent data structures is *serializability*. A concurrent data structure is serializable if for any concurrent execution of its operations, there exists an *equivalent* sequential execution of these operations. That is, each operation in the sequential execution has the same observable behavior (i.e., the same return value) as in the concurrent execution.

A number of stronger correctness criteria than serializability have been studied in the literature. These criteria impose additional constraints on the order of the operations in the equivalent sequential execution, relative to their order in the concurrent execution. One such criterion is *linearizability* [Herlihy and Wing, 1990], which we have used throughout this monograph. Intuitively, a concurrent object is linearizable if each of its operations appears to take place instantaneously at some point between its invocation and return point. This means that if an operation o_1 returns before the invocation of another operation o_2 in the concurrent execution, then o_1 must also appear before o_2 in the equivalent sequential execution. In particular, linearizability implies *sequential consistency*, another correctness criterion, which demands that operations in the sequential execution must occur in program order. Unlike sequential consistency, linearizability is compositional in the sense that the execution of a system composed of many concurrent objects is linearizable iff the projections of the execution to the operations of each individual object are linearizable. This means that one can reason about the correctness of individual objects in isolation, which simplifies the verification. Therefore, linearizability is often considered the gold standard for correctness of concurrent data structures. However, we note that some objects and operations are inherently nonlinearizable in which case weaker correctness notions must be considered [Emmi and Enea, 2019, Krishna et al., 2020a, Sergey et al., 2016].

Database transaction processing theory classically establishes the serializability of a set of concurrent transactions by showing that transactions satisfy a stronger condition known as *conflict-preserving serializability* [Bernstein et al., 1987]. Conflict-preserving serializability requires that transaction operations (reads and writes) can be safely reordered to arrive at a serial execution. Safe reordering does not change the operations involved but allows *non-conflicting* operations to be swapped. Two operations *conflict* if they access the same data and at least one is a write. This purely syntactic notion is sufficient for serializability but is too strong for concurrent search structure algorithms. For example, a search may read a node p then a split of a child c of p may move some keys from c to c' and change p accordingly, then the search may go to c and perhaps further to c'. There is no serial reordering of non-conflicting operations of this search and split in terms of reads and writes. Moreover, in no serial execution would a search go to both c and c', as Bob did in the library example of the introduction. Establishing correctness for concurrent search structures requires using the semantics of these structures and their operations. Conflict-preserving serializability is not enough.

For a more in-depth discussion of various correctness criteria for concurrent objects, we direct the interested reader to Herlihy and Shavit [2008].

Deductive verification of concurrent programs. In this monograph, we follow a deductive verification approach, which reduces reasoning about program correctness to discharging proof obligations expressed in a formal logic. Here, the program code is annotated with intermediate assertions that enable complex proof obligations to be mechanically decomposed into simpler ones according to program structure. This idea was pioneered by Turing [1949] and later cast into formal reasoning systems, starting with the development of (Floyd-)Hoare logic [Floyd, 1967, Hoare, 1969] and its mechanization due to Dijkstra [1975]. Since then, many formal systems have been proposed that enable compositional reasoning along dimensions of program complexity that are orthogonal to syntactic structure, including concurrency [Herlihy and Wing, 1990, Jones, 1983, Lamport, 1977, Owicki and Gries, 1976], data representation [Cousot and Cousot, 1977, Hoare, 1972], and time [Pnueli, 1977]. A number of textbooks provide detailed introductions to the relevant topics (see e.g., [Manna and Pnueli, 1995, Pierce et al., 2020, Winskel, 1993]).

Separation logic [Ishtiaq and O'Hearn, 2001, O'Hearn et al., 2001, Reynolds, 2002] is an extension of Hoare logic that was originally conceived to deal with the complexities imposed by mutable state. Separating conjunction and the accompanying frame rule enable "reasoning and specification to be confined to the cells that the program actually accesses" [O'Hearn et al., 2001]. O'Hearn soon realized that these reasoning principles naturally extend to concurrent programs that manipulate shared resources. This led to the development of concurrent separation logic [Brookes, 2004, O'Hearn, 2004], which has spawned a proliferation of logics that provide a sophisticated arsenal of modular reasoning techniques [Bornat et al., 2005, da Rocha Pinto et al., 2014, Dinsdale-Young et al., 2013, 2010, Feng et al., 2007, Gu et al., 2018, Heule et al., 2013, Jung et al., 2015, Nanevski et al., 2014, Raad et al., 2015, Vafeiadis and Parkinson, 2007, Xiong et al., 2017]. For a more comprehensive survey and discussion of the history of this development, we refer the reader to Brookes and O'Hearn [2016] and O'Hearn [2019].

We have formalized the verification of our template algorithms in the concurrent separation logic Iris [Jung et al., 2016, 2018, 2015, Krebbers et al., 2017a]. Our formalization particularly benefits from Iris' support for logically atomic triples [da Rocha Pinto et al., 2014, Frumin et al., 2018, Jacobs and Piessens, 2011, Jung et al., 2015] and user-definable resource algebras, which can capture nontrivial ghost state such as keysets and flow interfaces. However, we expect that our methodology can be replicated in other concurrent separation logics that support these features, such as FCSL [Sergey et al., 2015], which would also be useful if one wanted to extend the template-based approach to verifying nonlinearizable data structures [Sergey et al., 2016].

We make use of Iris' support for prophecy variables [Jung et al., 2020] to reason modularly about the non-local dynamic linearization points of searches in multicopy structures. Our proof discussed in Chapter 11 builds on the prophecy-based Iris proof of the RDCSS data structure from [Jung et al., 2020] and adapts it to a setting where an unbounded number of threads perform "helping". The idea of using prophecy variables to reason about non-fixed linearization points has also been explored in prior work building on other logics than Iris [Vafeiadis, 2008, Zhang et al., 2012], including situations that involve unbounded helping [Liang and Feng, 2013, Turon et al., 2013]. A possible alternative approach to using prophecies is to prove that the template-level atomic specification contextually refines the client-level atomic specification of multicopy structures using a relational program logic [Banerjee et al., 2016, Frumin et al., 2018, Liang and Feng, 2013].

Separating conjunction in Iris is affine, which means that it can be weakened by dropping one of its conjuncts. That is, the resources in an assertion can be "thrown away" without invalidating the assertion. In particular, this applies to resources that express ownership of allocated memory. Consequently, Iris cannot be used directly to reason about absence of memory leaks. Iron [Bizjak et al., 2019], a recent extension of Iris, allows proving absence of memory leaks in the context of manual memory management. In this monograph, we assume a garbage collected environment in our proofs. We argue below that this is a reasonable assumption, but relaxing this assumption is certainly attractive.

Another limitation of Iris is that it assumes that memory reads and writes are sequentially consistent. The logic lacks support for reasoning about the weaker consistency notions supported by modern hardware architectures. Recent work explores how to define separation logics for weak memory models [Svendsen et al., 2018]. For common models such as total store order, it is safe to assume sequential consistency if the program is data race free [Adve and Hill, 1993]. Data race freedom itself can be checked under the assumption of a sequentially consistent memory model [Bouajjani et al., 2013a]. All template algorithms considered in this monograph are lock-based and guarantee data race freedom, provided the locks are implemented correctly. Thus, reasoning about weak consistency can be confined to the verification of the lock implementation. Using similar reasoning, one can justify the use of a garbage collection semantics, which assumes that newly allocated memory is fresh, even when verifying programs that execute under manual memory management [Haziza et al., 2016].

There are numerous other formal proof systems that provide mechanisms for structuring the verification of complex concurrent programs and that do not build on separation logic. A common approach is to transform an abstract mathematical description of an algorithm to an efficient implementation using a sequence of refinement steps, each of which is formally justified by establishing a simulation relation between the model and its refinement [Abrial and

Hallerstede, 2007, Back, 1981, Back and Sere, 1989, Chandy and Misra, 1986, Kuppe et al., 2019, Lamport, 1994]. Other approaches use abstractions and reductions to combine primitive atomic operations into composite operations that are logically atomic [Elmas et al., 2010, Freund and Qadeer, 2004, Hawblitzel et al., 2015, Kragl and Qadeer, 2018, Kragl et al., 2020]. These approaches can be combined with techniques for reasoning about shared resources that specify frame conditions in classical logic using explicit ghost variables. Such techniques include, e.g., ownership-based techniques [Clarke et al., 1998, Cohen et al., 2010, Jacobs et al., 2005, Müller, 2001, Müller et al., 2006], dynamic frames [Kassios, 2006], and region logic [Banerjee et al., 2013]. Compared to separation logic, the explicit handling of frame reasoning incurs some overhead in terms of the complexity of specifications. Nevertheless, reasoning about these specifications can also be automated more directly using classical first-order theorem provers. Approaches such as implicit dynamic frames [Leino and Müller, 2009, Smans et al., 2009] and linear maps [Lahiri et al., 2011] aim to combine the best of both worlds (concise specifications and ease of automation). The verification methodology presented in this monograph is not inherently tied to separation logic. In fact, our implementation proofs are automated in a system based on implicit dynamic frames. A formal comparison between implicit dynamic frames and separation logic is given in Parkinson and Summers [2012].

Proof mechanization and automation.   Tool support for mechanizing and automating formal software verification ranges from implementations based on interactive proof assistants, which are versatile but require substantial user-guidance, to static analyzers and model checking tools, which are fully automated but more limited in the complexity of the programs and properties they can be applied to. In between lie semi-automated deductive verification tools. These tools require the user to annotate code with intermediate assertions, which are then used to generate verification conditions that are automatically discharged using decision procedures as implemented, e.g., in satisfiability modulo theories (SMT) solvers. A recent survey of state-of-the-art tools can be found in Hähnle and Huisman [2019].

We have mechanized the verification of the presented template algorithms for concurrent search structures in the Coq-based interactive proof mode of Iris [Krebbers et al., 2018, 2017b]. To verify that the implementations of the node-level helper functions assumed by the template algorithms satisfy their specifications, we have used the SMT-based deductive verifier GRASShopper [Piskac et al., 2014]. The presented approach is not bound to these specific tools. For example, we already have preliminary experience in automating flow-based proofs in the verification system Viper [Müller et al., 2017], which is based on the implicit dynamic frames formalism [Parkinson and Summers, 2012, Smans et al., 2009]. Moreover, many of the ideas and techniques proposed in this monograph can be used in other verification systems, including systems that are not based on separation logic such as Dafny [Leino, 2010, 2017] and CIVL [Hawblitzel et al., 2015].

Fully automated proofs of linearizability by static analysis and model checking have been mostly confined to simple list-based data structures [Abdulla et al., 2013, 2018, Amit et al.,

2007, Bouajjani et al., 2013b, 2015, 2017, Cerný et al., 2010, Dragoi et al., 2013, Henzinger et al., 2013, Liang and Feng, 2013, Vafeiadis, 2009, Zhu et al., 2015]. Recent work by Abdulla et al. [2018] shows how to automatically verify more complex structures such as concurrent skip lists that combine lists and arrays. Our work shows that it is possible to devise semi-automated techniques (in which one formulates useful invariants) that work over a broad class of diverse data structures. While full automation for such structures is still beyond the state of the art, we hope that ideas such as the keyset algebra and the accompanying inset flow will inform the design of future automated static analyzers.

Template algorithms.   Our work builds on the concurrent template algorithms for single-copy search structures of Shasha and Goodman [1988], extends it to multicopy structures, and develops a formal verification framework to verify such algorithms using state-of-the-art verification technology. Several other works present generic proof arguments for verifying concurrent traversals [Drachsler-Cohen et al., 2018, Feldman et al., 2018, 2020, O'Hearn et al., 2010]. These focus on lock-free search structures that have dynamic linearization points. However, they do not aim to decouple the reasoning about the thread synchronization mechanism from that of the underlying memory representation and data structure invariant.

Meyer and Wolff [2019, 2020] propose a technique that decouples the proof of data structure correctness from that of the underlying memory reclamation algorithm, allowing the correctness proof to be carried out under the assumption of garbage collection. The verified data structure implementations can then be composed with standard reclamation algorithms, e.g., based on epochs [Fraser, 2004] or hazard pointers [Michael, 2004]. It is a promising direction of future work to integrate these approaches and our technique in order to obtain verified data structures where the user can mix-and-match the synchronization technique, memory layout, and the memory reclamation algorithm.

14.2   FUTURE WORK

Combining the edgeset framework with the flow framework has allowed us to mechanically prove the safety of single-copy and multicopy search structures. We have shown such proofs for B-link trees, hash-tables, and log-structured merge (LSM) trees, among others. Here are some directions in which our work can be extended:

1. Extending the proof technique to prove liveness properties such as termination and other progress notions.

2. Generalizing the presented templates to cover other existing or potential search structure algorithms including those based on lock-free concurrent techniques.

We will now examine these directions in turn.

14.2.1   PROVING LIVENESS

Proving liveness properties such as progress or termination would involve strengthening the invariants we have used for our template algorithms. For example, the invariant that the keysets of any two nodes in a single-copy structure are disjoint was sufficient to prove partial correctness, but a liveness property like termination would require the stronger invariant that the set of keysets of all nodes cover the key space. This captures the high-level property that given any key k, there is some node in the structure that is responsible for k.

Proving that such an invariant is maintained would require us to prove that the underlying structure is, in some sense, acyclic. More precisely, we need to show that given any key k, its search path (i.e., the path obtained by starting at the root and following edges that have k in their edgeset) is acyclic. Note that none of the single-copy structure proofs in this monograph proved such an acyclicity property. For partial correctness, it is sufficient to show that if an operation on k operates on a node n then k is in n's keyset. Such an acyclicity invariant can be encoded by using the effectively-acyclic extension of the flow framework [Krishna et al., 2020c].

Liveness proofs usually proceed by defining a *ranking function*, a function from states of the structure to some well-ordered set. Once we have shown that the search structure is acyclic, we can define the ranking function of an operation as the cardinality of the search path. Acyclicity implies that the search path is a list, which means that every time the operation moves from one node to the next, the number of nodes in the search path decreases.

Note that when reasoning about liveness, it is very important to specify the environment under which the algorithm is operating. For instance, even in an acyclic structure, a search operation on k might never terminate if it is operating in an environment where maintenance operations are continually pre-empting the search operation and performing splits that increase k's search path. Standard assumptions about the environment or scheduler, such as fair scheduling or bounded interference, would be required to rule out such undesirable executions.

Formalizing liveness properties in separation logic would require us to use logics that formalize notions of fairness and ranking functions, such as TaDA-live [D'Osualdo et al., 2019]. We conjecture that our proof technique can be transferred to such logics without much technical difficulty because our flow-based encoding can be performed in any separation logic that supports user-defined resource algebras. There are also automated tools for proving that algorithms are non-blocking [Gotsman et al., 2009], and it would be interesting to see if we can incorporate our approach with these tools.

14.2.2   GENERALIZATIONS AND EXTENSIONS

In order to be able to use sequential reasoning in node-level operations, we have assumed the availability of a locking mechanism to protect accesses to individual nodes in the data structure graph. We can relax this assumption in various ways. For example, in the case of multicopy structures, each node is a single-copy search structure that may again consist of many nodes.

A natural generalization here is to allow concurrent updates to the nodes of these single-copy structures, e.g., when doing compaction.

Another direction is to consider data structures that do not use locks. Many lock-free algorithms have been proposed, e.g., [Harris, 2001, Levandoski et al., 2013] and analyzed [Bouajjani et al., 2013b, 2017, Chakraborty et al., 2015, Delbianco et al., 2017, Dodds et al., 2015, Frumin et al., 2018, Khyzha et al., 2017, Liang and Feng, 2013, O'Hearn et al., 2010, Zhu et al., 2015].

Very often, the template algorithms we propose continue to apply. For example, a lock-free list in which a thread prepends upserts to the beginning of the list by doing a CAS (compare-and-set) to the root pointer of the list (e.g., in a BW-tree [Levandoski et al., 2013]) is essentially a multicopy structure. Thus, it enjoys the same basic invariants as the log-structured merge tree: the logical value of a key is the value associated with the element closest to the head of the list. Properties like search recency still hold.

Further, an algorithm designer can often use compare-and-swap or the more general compare-and-set in a similar way to locks. Suppose, for example, that a thread records the state of some pointer at time t_1 and then does a compare-and-set based on that recorded state at time t_2. If the compare-and-set succeeds, then the pointer has not changed from t_1 to t_2. Note that if the thread had instead locked the pointer from t_1 to t_2, then the thread would enjoy the same guarantee.

In the prepending example above, "locking the root pointer" plays essentially the same role as "recording the address in the root pointer, constructing a node to prepend, and then doing a compare-and-swap on the root pointer". The advantage of locking is that there is no need to redo work, as there would be if the compare-and-swap fails. The advantage of the compare-and-swap is that lock-freedom avoids the possibility that a hung thread stops other threads from making progress, because that hung thread holds a lock.

The bottom line is that the *good state* conditions (i.e., the keysets of different nodes are disjoint, and the contents of each node is a subset of its keyset) still apply.

Consider, for example, the following simple (though very inefficient) single-copy lock-free concurrent tree algorithm.

- The root pointer is stored at a fixed location r. Each value of the address stored in that pointer is associated with the time of the last successful modification of the tree.

- A search at time t starts at r and follows the pointer to the copy of that tree structure present as of time t.

- Any modification (i) reads the address A stored in r which is the address of the root of the tree at that time; (ii) copies the existing tree structure and creates a whole new tree structure whose tree root node is at address A'[1]; (iii) performs its modification on the new tree; (iv) uses a compare-and-swap to check whether r still contains the address

---

[1]To ensure that tree root node addresses are never repeated, we might require that $A' > A$.

A and, if so, swaps it to address A'. If r no longer contains address A, the modification must start over.

- The linearization point of each modification is the time that r was successfully modified. The linearization point of every search is the time the search begins.

Of course, copying the entire search structure for every modification is extremely inefficient. An alternative is to (i) copy just the nodes in the path from the root to the leaf that is modified (say in a B-tree like structure), (ii) modify whichever nodes in that path that need to be modified, and then (iii) use a compare and swap to point to the new tree root. No nodes outside that path need be copied.

These methods work well with our proof methodology. The good state conditions are preserved by these algorithms and a thread that begins a search for k at time t and later visits a node n will enjoy the invariant that k is in the inset of n for the structure as of time t. Thus, the search's linearization point will be the time when the search begins, i.e., t.

Other lock-free algorithms such as the Harris list [Harris, 2001] (which maintains a list sorted by key) are more fine-grained and require new invariants, which we will address in future work.

14.3   CONCLUSION

We have presented a proof technique for concurrent search structures that (i) separates the reasoning about thread safety from memory safety using the edgeset framework; and (ii) allows local reasoning about global predicates using the flow framework.

We have demonstrated our technique by formalizing and verifying template algorithms using this technique, have shown how to derive verified implementations on specific data structures, and have thereby mechanized the proofs of linearizability and memory safety for a large class of concurrent search structures.

We believe that a decompositional flow-style approach can be used for other concurrent graph problems in communication networks and memory management. Those highly concurrent graph applications must preserve global properties such as reachability and connectivity through local operations.

Bibliography

Abadi, M. and Lamport, L. (1988). The existence of refinement mappings. In *Proc. of the 3rd Annual Symposium on Logic in Computer Science (LICS'88)*, pages 165–175, IEEE Computer Society, Edinburgh, Scotland, UK, July 5–8. DOI: 10.1109/lics.1988.5115 107

Abdulla, P. A., Haziza, F., Holík, L., Jonsson, B., and Rezine, A. (2013). An integrated specification and verification technique for highly concurrent data structures. In Piterman, N. and Smolka, S. A., Eds., *Tools and Algorithms for the Construction and Analysis of Systems—19th International Conference, TACAS, Held as Part of the European Joint Conferences on Theory and Practice of Software, ETAPS, Proceedings*, vol. 7795 of *Lecture Notes in Computer Science*, pages 324–338, Springer, Rome, Italy, March 16–24. DOI: 10.1007/978-3-642-36742-7_23 152

Abdulla, P. A., Jonsson, B., and Trinh, C. Q. (2018). Fragment abstraction for concurrent shape analysis. In Ahmed, A., Ed., *Programming Languages and Systems—27th European Symposium on Programming, ESOP, Held as Part of the European Joint Conferences on Theory and Practice of Software, ETAPS, Proceedings*, vol. 10801 of *Lecture Notes in Computer Science*, pages 442–471, Springer, Thessaloniki, Greece, April 14–20. DOI: 10.1007/978-3-319-89884-1_16 152, 153

Abrial, J. and Hallerstede, S. (2007). Refinement, decomposition, and instantiation of discrete models: Application to event-b. *Fundam. Informaticae*, 77(1–2):1–28. 151

Adve, S. V. and Hill, M. D. (1993). A unified formalization of four shared-memory models. *IEEE Trans. Parall. Distrib. Syst.*, 4(6):613–624. DOI: 10.1109/71.242161 151

Amit, D., Rinetzky, N., Reps, T. W., Sagiv, M., and Yahav, E. (2007). Comparison under abstraction for verifying linearizability. In Damm, W. and Hermanns, H., Eds., *Computer Aided Verification, 19th International Conference, CAV, Proceedings*, vol. 4590 of *Lecture Notes in Computer Science*, pages 477–490, Springer, Berlin, Germany, July 3–7. DOI: 10.1007/978-3-540-73368-3_49 152

Apache Software Foundation (2020a). Apache cassandra. https://cassandra.apache.org/ 97, 134

Apache Software Foundation (2020b). Apache HBase. https://hbase.apache.org/ 101

Back, R. (1981). On correct refinement of programs. *J. Comput. Syst. Sci.*, 23(1):49–68. DOI: 10.1016/0022-0000(81)90005-2 3, 152

Back, R. and Sere, K. (1989). Stepwise refinement of parallel algorithms. *Sci. Comput. Program.*, 13(1):133–180. DOI: 10.1016/0167-6423(90)90069-p 152

Ball, T., Levin, V., and Rajamani, S. K. (2011). A decade of software model checking with SLAM. *Commun. ACM*, 54(7):68–76. DOI: 10.1145/1965724.1965743 2

Banerjee, A., Naumann, D. A., and Nikouei, M. (2016). Relational logic with framing and hypotheses. In Lal, A., Akshay, S., Saurabh, S., and Sen, S., Eds., *36th IARCS Annual Conference on Foundations of Software Technology and Theoretical Computer Science, FSTTCS*, vol. 65 of *LIPIcs*, pages 11:1–11:16, Schloss Dagstuhl – Leibniz-Zentrum für Informatik, Chennai, India, December 13–15. 151

Banerjee, A., Naumann, D. A., and Rosenberg, S. (2013). Local reasoning for global invariants, part I: Region logic. *J. ACM*, 60(3):18:1–18:56. DOI: 10.1145/2487241.2485982 3, 152

Bansal, K., Reynolds, A., King, T., Barrett, C. W., and Wies, T. (2015). Deciding local theory extensions via e-matching. In Kroening, D. and Pasareanu, C. S., Eds., *Computer Aided Verification—27th International Conference, CAV, Proceedings, Part II*, vol. 9207 of *Lecture Notes in Computer Science*, pages 87–105, Springer, San Francisco, CA, July 18–24. DOI: 10.1007/978-3-319-21668-3_6 147

Bernstein, P. A., Hadzilacos, V., and Goodman, N. (1987). *Concurrency Control and Recovery in Database Systems*. Addison-Wesley. DOI: 10.1145/356842.356846 31, 94, 149

Bhargavan, K., Bond, B., Delignat-Lavaud, A., Fournet, C., Hawblitzel, C., Hriţcu, C., Ishtiaq, S., Kohlweiss, M., Leino, R., Lorch, J. R., Maillard, K., Pan, J., Parno, B., Protzenko, J., Ramananandro, T., Rane, A., Rastogi, A., Swamy, N., Thompson, L., Wang, P., Béguelin, S. Z., and Zinzindohoue, J. K. (2017). Everest: Towards a verified, drop-in replacement of HTTPS. In Lerner, B. S., Bodík, R., and Krishnamurthi, S., Eds., *2nd Summit on Advances in Programming Languages, SNAPL*, vol. 71 of *LIPIcs*, pages 1:1–1:12, Schloss Dagstuhl – Leibniz-Zentrum für Informatik, Asilomar, CA, May 7–10. DOI: 10.1145/781131.781153 2

Bizjak, A., Gratzer, D., Krebbers, R., and Birkedal, L. (2019). Iron: Managing obligations in higher-order concurrent separation logic. *Proc. ACM Program. Lang.*, 3(POPL):65:1–65:30. DOI: 10.1145/3290378 17, 151

Blanchet, B., Cousot, P., Cousot, R., Feret, J., Mauborgne, L., Miné, A., Monniaux, D., and Rival, X. (2003). A static analyzer for large safety-critical software. In Cytron, R. and Gupta, R., Eds., *Proc. of the ACM SIGPLAN Conference on Programming Language Design and Implementation*, pages 196–207, San Diego, CA, June 9–11. 2

Bornat, R., Calcagno, C., and Yang, H. (2005). Variables as resource in separation logic. In Escardó, M. H., Jung, A., and Mislove, M. W., Eds., *Proc. of the 21st Annual Conference on Mathematical Foundations of Programming Semantics, MFPS*, vol. 155 of *Electronic Notes in Theoretical Computer Science*, pages 247–276, Elsevier, Birmingham, UK, May 18–21. DOI: 10.1016/j.entcs.2005.11.059 150

Bouajjani, A., Derevenetc, E., and Meyer, R. (2013a). Checking and enforcing robustness against TSO. In Felleisen, M. and Gardner, P., Eds., *Programming Languages and Systems— 22nd European Symposium on Programming, ESOP, Held as Part of the European Joint Conferences on Theory and Practice of Software, ETAPS, Proceedings*, vol. 7792 of *Lecture Notes in Computer Science*, pages 533–553, Springer, Rome, Italy, March 16–24. DOI: 10.1007/978-3-642-37036-6_29 151

Bouajjani, A., Emmi, M., Enea, C., and Hamza, J. (2013b). Verifying concurrent programs against sequential specifications. In Felleisen, M. and Gardner, P., Eds., *Programming Languages and Systems—22nd European Symposium on Programming, ESOP, Held as Part of the European Joint Conferences on Theory and Practice of Software, ETAPS, Proceedings*, vol. 7792 of *Lecture Notes in Computer Science*, pages 290–309, Springer, Rome, Italy, March 16–24. DOI: 10.1007/978-3-642-37036-6_17 153, 155

Bouajjani, A., Emmi, M., Enea, C., and Hamza, J. (2015). On reducing linearizability to state reachability. In Halldórsson, M. M., Iwama, K., Kobayashi, N., and Speckmann, B., Eds., *Automata, Languages, and Programming—42nd International Colloquium, ICALP, Proceedings, Part II*, vol. 9135 of *Lecture Notes in Computer Science*, pages 95–107, Springer, Kyoto, Japan, July 6–10. DOI: 10.1007/978-3-662-47666-6_8 153

Bouajjani, A., Emmi, M., Enea, C., and Mutluergil, S. O. (2017). Proving linearizability using forward simulations. In Majumdar, R. and Kuncak, V., Eds., *Computer Aided Verification— 29th International Conference, CAV, Proceedings, Part II*, vol. 10427 of *Lecture Notes in Computer Science*, pages 542–563, Springer, Heidelberg, Germany, July 24–28. DOI: 10.1007/978-3-319-63390-9_28 153, 155

Brookes, S. (2007). A semantics for concurrent separation logic. *Theor. Comput. Sci.*, 375(1–3):227–270. DOI: 10.1016/j.tcs.2006.12.034 3

Brookes, S. and O'Hearn, P. W. (2016). Concurrent separation logic. *ACM SIGLOG News*, 3(3):47–65. DOI: 10.1145/2984450.2984457 3, 150

Brookes, S. D. (2004). A semantics for concurrent separation logic. In Gardner, P. and Yoshida, N., Eds., *CONCUR—Concurrency Theory, 15th International Conference, Proceedings*, vol. 3170 of *Lecture Notes in Computer Science*, pages 16–34, Springer, London, UK, August 31–September 3. DOI: 10.1007/978-3-540-28644-8_2 150

Burckhardt, S., Alur, R., and Martin, M. M. K. (2007). Checkfence: Checking consistency of concurrent data types on relaxed memory models. In Ferrante, J. and McKinley, K. S., Eds., *Proc. of the ACM SIGPLAN Conference on Programming Language Design and Implementation*, pages 12–21, San Diego, CA, June 10–13. DOI: 10.1145/1250734.1250737 1

Cerný, P., Radhakrishna, A., Zufferey, D., Chaudhuri, S., and Alur, R. (2010). Model checking of linearizability of concurrent list implementations. In Touili, T., Cook, B., and Jackson, P. B., Eds., *Computer Aided Verification, 22nd International Conference, CAV, Proceedings*, vol. 6174 of *Lecture Notes in Computer Science*, pages 465–479, Springer, Edinburgh, UK, July 15–19. DOI: 10.1007/978-3-642-14295-6_41 153

Chajed, T., Jung, R., and Tassarotti, J. (2021). Iris tutorial. https://gitlab.mpi-sws.org/iris/tutorial-popl21 17, 27, 37

Chakraborty, S., Henzinger, T. A., Sezgin, A., and Vafeiadis, V. (2015). Aspect-oriented linearizability proofs. *Log. Meth. Comput. Sci.*, 11(1). DOI: 10.2168/lmcs-11(1:20)2015 155

Chandy, K. M. and Misra, J. (1986). An example of stepwise refinement of distributed programs: Quiescence detection. *ACM Trans. Program. Lang. Syst.*, 8(3):326–343. DOI: 10.1145/5956.5958 152

Chong, N., Cook, B., Kallas, K., Khazem, K., Monteiro, F. R., Schwartz-Narbonne, D., Tasiran, S., Tautschnig, M., and Tuttle, M. R. (2020). Code-level model checking in the software development workflow. In Rothermel, G. and Bae, D., Eds., *ICSE-SEIP: 42nd International Conference on Software Engineering, Software Engineering in Practice*, pages 11–20, ACM, Seoul, South Korea, June 27–July 19. DOI: 10.1145/3377813.3381347 2

Clarke, D. G., Potter, J., and Noble, J. (1998). Ownership types for flexible alias protection. In Freeman-Benson, B. N. and Chambers, C., Eds., *Proc. of the ACM SIGPLAN Conference on Object-Oriented Programming Systems, Languages and Applications (OOPSLA'98)*, pages 48–64, Vancouver, British Columbia, Canada, October 18–22. DOI: 10.1145/286936.286947 152

Cohen, E., Moskal, M., Schulte, W., and Tobies, S. (2010). Local verification of global invariants in concurrent programs. In Touili, T., Cook, B., and Jackson, P. B., Eds., *Computer Aided Verification, 22nd International Conference, CAV, Proceedings*, vol. 6174 of *Lecture Notes in Computer Science*, pages 480–494, Springer, Edinburgh, UK, July 15–19. DOI: 10.1007/978-3-642-14295-6_42 152

Cousot, P. and Cousot, R. (1977). Abstract interpretation: A unified lattice model for static analysis of programs by construction or approximation of fixpoints. In Graham, R. M., Harrison, M. A., and Sethi, R., Eds., *Conference Record of the 4th ACM Symposium on Principles of Programming Languages*, pages 238–252, Los Angeles, CA. DOI: 10.1145/512950.512973 150 da Rocha Pinto, P., Dinsdale-Young, T., Dodds, M., Gardner, P., and Wheelhouse, M. J. (2011). A simple abstraction for complex concurrent indexes. In Lopes, C. V. and Fisher, K., Eds., *Proc. of the 26th Annual ACM SIGPLAN Conference on Object-Oriented Programming, Systems, Languages, and Applications, OOPSLA, part of SPLASH*, pages 845–864, Portland, OR, October 22–27. DOI: 10.1145/2048066.2048131 3, 4

da Rocha Pinto, P., Dinsdale-Young, T., and Gardner, P. (2014). Tada: A logic for time and data abstraction. In Jones, R. E., Ed., *ECOOP—Object-Oriented Programming—28th European Conference, Proceedings*, vol. 8586 of *Lecture Notes in Computer Science*, pages 207–231, Springer, Uppsala, Sweden, July 28–August 1. DOI: 10.1007/978-3-662-44202-9_9 3, 5, 31, 150

Delbianco, G. A., Sergey, I., Nanevski, A., and Banerjee, A. (2017). Concurrent data structures linked in time. In Müller, P., Ed., *31st European Conference on Object-Oriented Programming, ECOOP*, vol. 74 of *LIPIcs*, pages 8:1–8:30, Schloss Dagstuhl – Leibniz-Zentrum für Informatik, Barcelona, Spain, June 19–23. 155

Dijkstra, E. W. (1968). A constructive approach to the problem of program correctness. *BIT*, 8:174–186. DOI: 10.1007/bf01933419 3

Dijkstra, E. W. (1975). Guarded commands, nondeterminacy and formal derivation of programs. *Commun. ACM*, 18(8):453–457. DOI: 10.1145/360933.360975 150

Dinsdale-Young, T., Birkedal, L., Gardner, P., Parkinson, M. J., and Yang, H. (2013). Views: Compositional reasoning for concurrent programs. In Giacobazzi, R. and Cousot, R., Eds., *The 40th Annual ACM SIGPLAN-SIGACT Symposium on Principles of Programming Languages, POPL'13*, pages 287–300, Rome, Italy, January 23–25. DOI: 10.1145/2429069.2429104 3, 150

Dinsdale-Young, T., Dodds, M., Gardner, P., Parkinson, M. J., and Vafeiadis, V. (2010). Concurrent abstract predicates. In D'Hondt, T., Ed., *ECOOP—Object-Oriented Programming, 24th European Conference, Proceedings*, vol. 6183 of *Lecture Notes in Computer Science*, pages 504–528, Springer, Maribor, Slovenia, June 21–25. DOI: 10.1007/978-3-642-14107-2_24 3, 150

Distefano, D., Fähndrich, M., Logozzo, F., and O'Hearn, P. W. (2019). Scaling static analyses at Facebook. *Commun. ACM*, 62(8):62–70. DOI: 10.1145/3338112 2

Dodds, M., Haas, A., and Kirsch, C. M. (2015). A scalable, correct time-stamped stack. In Rajamani, S. K. and Walker, D., Eds., *Proc. of the 42nd Annual ACM SIGPLAN-SIGACT Symposium on Principles of Programming Languages, POPL*, pages 233–246, Mumbai, India, January 15–17. DOI: 10.1145/2775051.2676963 155

Dodds, M., Jagannathan, S., Parkinson, M. J., Svendsen, K., and Birkedal, L. (2016). Verifying custom synchronization constructs using higher-order separation logic. *ACM Trans. Program. Lang. Syst.*, 38(2):4:1–4:72. DOI: 10.1145/2818638 3

D'Osualdo, E., Farzan, A., Gardner, P., and Sutherland, J. (2019). Tada live: Compositional reasoning for termination of fine-grained concurrent programs. *CoRR*. 154

Drachsler-Cohen, D., Vechev, M. T., and Yahav, E. (2018). Practical concurrent traversals in search trees. In Krall, A. and Gross, T. R., Eds., *Proc. of the 23rd ACM SIGPLAN Symposium on Principles and Practice of Parallel Programming, PPoPP*, pages 207–218, Vienna, Austria, February 24–28. DOI: 10.1145/3178487.3178503 153

Dragoi, C., Gupta, A., and Henzinger, T. A. (2013). Automatic linearizability proofs of concurrent objects with cooperating updates. In Sharygina, N. and Veith, H., Eds., *Computer Aided Verification—25th International Conference, CAV, Proceedings*, vol. 8044 of *Lecture Notes in Computer Science*, pages 174–190, Springer, Saint Petersburg, Russia, July 13–19. DOI: 10.1007/978-3-642-39799-8_11 153

Elmas, T., Qadeer, S., Sezgin, A., Subasi, O., and Tasiran, S. (2010). Simplifying linearizability proofs with reduction and abstraction. In Esparza, J. and Majumdar, R., Eds., *Tools and Algorithms for the Construction and Analysis of Systems, 16th International Conference, TACAS, Held as Part of the Joint European Conferences on Theory and Practice of Software, ETAPS, Proceedings*, vol. 6015 of *Lecture Notes in Computer Science*, pages 296–311, Springer, Paphos, Cyprus, March 20–28. DOI: 10.1007/978-3-642-12002-2_25 152

Emmi, M. and Enea, C. (2019). Weak-consistency specification via visibility relaxation. *Proc. ACM Program. Lang.*, 3(POPL):60:1–60:28. DOI: 10.1145/3290373 149

Facebook (2020). RocksDB. https://rocksdb.org/ 101

Feldman, Y. M. Y., Enea, C., Morrison, A., Rinetzky, N., and Shoham, S. (2018). Order out of chaos: Proving linearizability using local views. In Schmid, U. and Widder, J., Eds., *32nd International Symposium on Distributed Computing, DISC*, vol. 121 of *LIPIcs*, pages 23:1–23:21, Schloss Dagstuhl – Leibniz-Zentrum für Informatik, New Orleans, LA, October 15–19. 153

Feldman, Y. M. Y., Khyzha, A., Enea, C., Morrison, A., Nanevski, A., Rinetzky, N., and Shoham, S. (2020). Proving highly-concurrent traversals correct. *Proc. ACM Program. Lang.*, 4(OOPSLA):128:1–128:29. DOI: 10.1145/3428196 153

Feng, X., Ferreira, R., and Shao, Z. (2007). On the relationship between concurrent separation logic and assume-guarantee reasoning. In Nicola, R. D., Ed., *Programming Languages and Systems, 16th European Symposium on Programming, ESOP, Held as Part of the Joint European Conferences on Theory and Practices of Software, ETAPS, Proceedings*, vol. 4421 of *Lecture*

Notes in Computer Science, pages 173–188, Springer, Braga, Portugal, March 24–April 1. DOI: 10.1007/978-3-540-71316-6_13 3, 150

Filipovic, I., O'Hearn, P. W., Rinetzky, N., and Yang, H. (2009). Abstraction for concurrent objects. In Castagna, G., Ed., *Programming Languages and Systems, 18th European Symposium on Programming, ESOP 2009, Held as Part of the Joint European Conferences on Theory and Practice of Software, ETAPS 2009, York, UK, March 22–29, 2009. Proceedings*, vol. 5502 of *Lecture Notes in Computer Science*, pages 252–266, Springer, DOI: 10.1007/978-3-642-00590-9_19 31

Floyd, R. W. (1967). Assigning meanings to programs. *Proc. of Symposium on Applied Mathematics*, 19:19–32. DOI: 10.1090/psapm/019/0235771 150

Fraser, K. (2004). Practical lock-freedom. Ph.D. thesis, University of Cambridge, UK. 153

Freund, S. N. and Qadeer, S. (2004). Checking concise specifications for multithreaded software. *J. Object Technol.*, 3(6):81–101. DOI: 10.5381/jot.2004.3.6.a4 152

Frumin, D., Krebbers, R., and Birkedal, L. (2018). Reloc: A mechanised relational logic for fine-grained concurrency. In Dawar, A. and Grädel, E., Eds., *Proc. of the 33rd Annual ACM/IEEE Symposium on Logic in Computer Science, LICS*, pages 442–451, Oxford, UK, July 09–12. DOI: 10.1145/3209108.3209174 150, 151, 155

Fu, M., Li, Y., Feng, X., Shao, Z., and Zhang, Y. (2010). Reasoning about optimistic concurrency using a program logic for history. In Gastin, P. and Laroussinie, F., Eds., *CONCUR 2010—Concurrency Theory, 21th International Conference, CONCUR, Proceedings*, vol. 6269 of *Lecture Notes in Computer Science*, pages 388–402, Springer, Paris, France, August 31–September 3. DOI: 10.1007/978-3-642-15375-4_27 3

Google (2020). LevelDB. https://github.com/google/leveldb 97, 101, 134, 147

Gotsman, A., Cook, B., Parkinson, M. J., and Vafeiadis, V. (2009). Proving that non-blocking algorithms don't block. In Shao, Z. and Pierce, B. C., Eds., *Proc. of the 36th ACM SIGPLAN-SIGACT Symposium on Principles of Programming Languages, POPL*, pages 16–28, Savannah, GA, January 21–23. DOI: 10.1145/1594834.1480886 154

Gu, R., Shao, Z., Kim, J., Wu, X. N., Koenig, J., Sjöberg, V., Chen, H., Costanzo, D., and Ramananandro, T. (2018). Certified concurrent abstraction layers. In Foster, J. S. and Grossman, D., Eds., *Proc. of the 39th ACM SIGPLAN Conference on Programming Language Design and Implementation, PLDI*, pages 646–661, Philadelphia, PA, June 18–22. DOI: 10.1145/3296979.3192381 150

Hähnle, R. and Huisman, M. (2019).   Deductive software verification: From pen-and-paper proofs to industrial tools.   In Steffen, B. and Woeginger, G. J., Eds., *Computing and Software Science—State-of-the-Art and Perspectives*, vol. 10000 of *Lecture Notes in Computer Science*, pages 345–373, Springer. DOI: 10.1007/978-3-319-91908-9_18 152

Harris, T. L. (2001).   A pragmatic implementation of non-blocking linked-lists.   In Welch, J. L., Ed., *Distributed Computing, 15th International Conference, DISC, Proceedings*, vol. 2180 of *Lecture Notes in Computer Science*, pages 300–314, Springer, Lisbon, Portugal, October 3–5. DOI: 10.1007/3-540-45414-4_21 155, 156

Hawblitzel, C., Petrank, E., Qadeer, S., and Tasiran, S. (2015).   Automated and modular refinement reasoning for concurrent programs.   In Kroening, D. and Pasareanu, C. S., Eds., *Computer Aided Verification—27th International Conference, CAV, Proceedings, Part II*, vol. 9207 of *Lecture Notes in Computer Science*, pages 449–465, Springer, San Francisco, CA, July 18–24. DOI: 10.1007/978-3-319-21668-3_26 152

Haziza, F., Holík, L., Meyer, R., and Wolff, S. (2016).   Pointer race freedom.   In Jobstmann, B. and Leino, K. R. M., Eds., *Verification, Model Checking, and Abstract Interpretation—17th International Conference, VMCAI, Proceedings*, vol. 9583 of *Lecture Notes in Computer Science*, pages 393–412, Springer, St. Petersburg, FL, January 17–19. DOI: 10.1007/978-3-662-49122-5_19 151

Henzinger, T. A., Sezgin, A., and Vafeiadis, V. (2013).   Aspect-oriented linearizability proofs.   In D'Argenio, P. R. and Melgratti, H. C., Eds., *CONCUR—Concurrency Theory—24th International Conference, Proceedings*, vol. 8052 of *Lecture Notes in Computer Science*, pages 242–256, Springer, Buenos Aires, Argentina, August 27–30. DOI: 10.1007/978-3-642-40184-8_18 153

Herlihy, M. and Shavit, N. (2008). *The Art of Multiprocessor Programming*. Morgan Kaufmann. 1, 150

Herlihy, M. and Wing, J. M. (1990).   Linearizability: A correctness condition for concurrent objects. *ACM Trans. Program. Lang. Syst.*, 12(3):463–492. DOI: 10.1145/78969.78972 3, 5, 31, 149, 150

Heule, S., Leino, K. R. M., Müller, P., and Summers, A. J. (2013).   Abstract read permissions: Fractional permissions without the fractions.   In Giacobazzi, R., Berdine, J., and Mastroeni, I., Eds., *Verification, Model Checking, and Abstract Interpretation, 14th International Conference, VMCAI, Proceedings*, vol. 7737 of *Lecture Notes in Computer Science*, pages 315–334, Springer, Rome, Italy, January 20–22. DOI: 10.1007/978-3-642-35873-9_20 150

Hoare, C. A. R. (1969).   An axiomatic basis for computer programming. *Commun. ACM*, 12(10):576–580. DOI: 10.1145/363235.363259 3, 17, 18, 150

Hoare, C. A. R. (1972). Proof of correctness of data representations. *Acta Informatica*, 1:271–281. DOI: 10.1007/bf00289507 3, 26, 150

Hoare, C. A. R. (1974). Monitors: An operating system structuring concept. *Commun. ACM*, 17(10):549–557. DOI: 10.1145/355620.361161 38

Iris Team (2020). The Iris 3.3 documentation. https://plv.mpi-sws.org/iris/appendix-3.3.pdf 17, 23, 53

Ishtiaq, S. S. and O'Hearn, P. W. (2001). BI as an assertion language for mutable data structures. In Hankin, C. and Schmidt, D., Eds., *Conference Record of POPL: The 28th ACM SIGPLAN-SIGACT Symposium on Principles of Programming Languages*, pages 14–26, London, UK, January 17–19. DOI: 10.1145/360204.375719 4, 150

Jacobs, B. and Piessens, F. (2011). Expressive modular fine-grained concurrency specification. In Ball, T. and Sagiv, M., Eds., *Proc. of the 38th ACM SIGPLAN-SIGACT Symposium on Principles of Programming Languages, POPL*, pages 271–282, Austin, TX, January 26–28. DOI: 10.1145/1925844.1926417 150

Jacobs, B., Piessens, F., Leino, K. R. M., and Schulte, W. (2005). Safe concurrency for aggregate objects with invariants. In Aichernig, B. K. and Beckert, B., Eds., *3rd IEEE International Conference on Software Engineering and Formal Methods (SEFM)*, pages 137–147, IEEE Computer Society, Koblenz, Germany, September 7–9. DOI: 10.1109/sefm.2005.39 152

Jonathan Ellis (2011). Leveled compaction in Apache Cassandra. https://www.datastax.com/blog/2011/10/leveled-compaction-apache-cassandra 134

Jones, C. B. (1983). Specification and design of (parallel) programs. In Mason, R. E. A., Ed., *Information Processing, Proceedings of the IFIP 9th World Computer Congress*, pages 321–332, North-Holland/IFIP, Paris, France, September 19–23. 3, 150

Jung, R., Krebbers, R., Birkedal, L., and Dreyer, D. (2016). Higher-order ghost state. In Garrigue, J., Keller, G., and Sumii, E., Eds., *Proc. of the 21st ACM SIGPLAN International Conference on Functional Programming, ICFP*, pages 256–269, Nara, Japan, September 18–22. DOI: 10.1145/3022670.2951943 5, 150

Jung, R., Krebbers, R., Jourdan, J., Bizjak, A., Birkedal, L., and Dreyer, D. (2018). Iris from the ground up: A modular foundation for higher-order concurrent separation logic. *J. Funct. Program.*, 28:e20. DOI: 10.1017/s0956796818000151 5, 8, 17, 37, 53, 150

Jung, R., Lepigre, R., Parthasarathy, G., Rapoport, M., Timany, A., Dreyer, D., and Jacobs, B. (2020). The future is ours: Prophecy variables in separation logic. *Proc. ACM Program. Lang.*, 4(POPL):45:1–45:32. DOI: 10.1145/3371113 5, 31, 107, 109, 151

Jung, R., Swasey, D., Sieczkowski, F., Svendsen, K., Turon, A., Birkedal, L., and Dreyer, D. (2015). Iris: Monoids and invariants as an orthogonal basis for concurrent reasoning. In Rajamani, S. K. and Walker, D., Eds., *Proc. of the 42nd Annual ACM SIGPLAN-SIGACT Symposium on Principles of Programming Languages, POPL*, pages 637–650, Mumbai, India, January 15–17. DOI: 10.1145/2676726.2676980 3, 5, 8, 31, 145, 150

Kassios, I. T. (2006). Dynamic frames: Support for framing, dependencies and sharing without restrictions. In Misra, J., Nipkow, T., and Sekerinski, E., Eds., *FM: Formal Methods, 14th International Symposium on Formal Methods, Hamilton, Proceedings*, vol. 4085 of *Lecture Notes in Computer Science*, pages 268–283, Springer, Canada, August 21–27. DOI: 10.1007/11813040_19 3, 152

Khyzha, A., Dodds, M., Gotsman, A., and Parkinson, M. J. (2017). Proving linearizability using partial orders. In Yang, H., Ed., *Programming Languages and Systems—26th European Symposium on Programming, ESOP, Held as Part of the European Joint Conferences on Theory and Practice of Software, ETAPS, Proceedings*, vol. 10201 of *Lecture Notes in Computer Science*, pages 639–667, Springer, Uppsala, Sweden, April 22–29. DOI: 10.1007/978-3-662-54434-1_24 155

Klein, G., Andronick, J., Elphinstone, K., Heiser, G., Cock, D., Derrin, P., Elkaduwe, D., Engelhardt, K., Kolanski, R., Norrish, M., Sewell, T., Tuch, H., and Winwood, S. (2010). sel4: Formal verification of an operating-system kernel. *Commun. ACM*, 53(6):107–115. DOI: 10.1145/1743546.1743574 2

Kragl, B. and Qadeer, S. (2018). Layered concurrent programs. In Chockler, H. and Weissenbacher, G., Eds., *Computer Aided Verification—30th International Conference, CAV, Held as Part of the Federated Logic Conference, FloC, Proceedings, Part I*, vol. 10981 of *Lecture Notes in Computer Science*, pages 79–102, Springer, Oxford, UK, July 14–17. DOI: 10.1007/978-3-319-96145-3_5 152

Kragl, B., Qadeer, S., and Henzinger, T. A. (2020). Refinement for structured concurrent programs. In Lahiri, S. K. and Wang, C., Eds., *Computer Aided Verification—32nd International Conference, CAV, Proceedings, Part I*, vol. 12224 of *Lecture Notes in Computer Science*, pages 275–298, Springer, Los Angeles, CA, July 21–24. DOI: 10.1007/978-3-030-53288-8_14 152

Krebbers, R., Jourdan, J., Jung, R., Tassarotti, J., Kaiser, J., Timany, A., Charguéraud, A., and Dreyer, D. (2018). Mosel: a general, extensible modal framework for interactive proofs in separation logic. *Proc. ACM Program. Lang.*, 2(ICFP):77:1–77:30. DOI: 10.1145/3236772 5, 145, 152

Krebbers, R., Jung, R., Bizjak, A., Jourdan, J., Dreyer, D., and Birkedal, L. (2017a). The essence of higher-order concurrent separation logic. In Yang, H., Ed., *Programming Lan-*

guages and Systems—26th European Symposium on Programming, ESOP, Held as Part of the European Joint Conferences on Theory and Practice of Software, ETAPS, Proceedings, vol. 10201 of *Lecture Notes in Computer Science*, pages 696–723, Springer, Uppsala, Sweden, April 22–29. DOI: 10.1007/978-3-662-54434-1_26 5, 150

Krebbers, R., Timany, A., and Birkedal, L. (2017b). Interactive proofs in higher-order concurrent separation logic. In Castagna, G. and Gordon, A. D., Eds., *Proc. of the 44th ACM SIGPLAN Symposium on Principles of Programming Languages, POPL*, pages 205–217, Paris, France, January 18–20. DOI: 10.1145/3093333.3009855 5, 145, 152

Krishna, S., Emmi, M., Enea, C., and Jovanovic, D. (2020a). Verifying visibility-based weak consistency. In Müller, P., Ed., *Programming Languages and Systems—29th European Symposium on Programming, ESOP, Held as Part of the European Joint Conferences on Theory and Practice of Software, ETAPS, Proceedings*, vol. 12075 of *Lecture Notes in Computer Science*, pages 280–307, Springer, Dublin, Ireland, April 25–30. DOI: 10.1007/978-3-030-44914-8_11 149

Krishna, S., Patel, N., Shasha, D. E., and Wies, T. (2020b). Verifying concurrent search structure templates. In Donaldson, A. F. and Torlak, E., Eds., *Proc. of the 41st ACM SIGPLAN International Conference on Programming Language Design and Implementation, PLDI*, pages 181–196, London, UK, June 15–20. DOI: 10.1145/3385412.3386029 8, 67

Krishna, S., Shasha, D. E., and Wies, T. (2018). Go with the flow: Compositional abstractions for concurrent data structures. *Proc. ACM Program. Lang.*, 2(POPL):37:1–37:31. DOI: 10.1145/3158125 5, 67

Krishna, S., Summers, A. J., and Wies, T. (2020c). Local reasoning for global graph properties. In Müller, P., Ed., *Programming Languages and Systems—29th European Symposium on Programming, ESOP, Held as Part of the European Joint Conferences on Theory and Practice of Software, ETAPS, Proceedings*, vol. 12075 of *Lecture Notes in Computer Science*, pages 308–335, Springer, Dublin, Ireland, April 25–30. DOI: 10.1007/978-3-030-44914-8_12 5, 8, 67, 71, 154

Kuppe, M. A., Lamport, L., and Ricketts, D. (2019). The TLA+ toolbox. In Monahan, R., Prevosto, V., and Proença, J., Eds., *Proc. of the 5th Workshop on Formal Integrated Development Environment, F-IDE@FM*, vol. 310 of *EPTCS*, pages 50–62, Porto, Portugal, October 7. DOI: 10.4204/eptcs.310.6 152

Lahiri, S. K., Qadeer, S., and Walker, D. (2011). Linear maps. In Jhala, R. and Swierstra, W., Eds., *Proc. of the 5th ACM Workshop Programming Languages meets Program Verification, PLPV*, pages 3–14, Austin, TX, January 29. DOI: 10.1145/1929529.1929531 152

Lamport, L. (1977). Proving the correctness of multiprocess programs. *IEEE Trans. Software Eng.*, 3(2):125–143. DOI: 10.1109/tse.1977.229904 150

Lamport, L. (1994). The temporal logic of actions. *ACM Trans. Program. Lang. Syst.*, 16(3):872–923. DOI: 10.1145/177492.177726 152

Leino, K. R. M. (2010). Dafny: An automatic program verifier for functional correctness. In Clarke, E. M. and Voronkov, A., Eds., *Logic for Programming, Artificial Intelligence, and Reasoning—16th International Conference, LPAR-16, Revised Selected Papers*, vol. 6355 of *Lecture Notes in Computer Science*, pages 348–370, Springer, Dakar, Senegal, April 25–May 1. DOI: 10.1007/978-3-642-17511-4_20 152

Leino, K. R. M. (2017). Modeling concurrency in dafny. In Bowen, J. P., Liu, Z., and Zhang, Z., Eds., *Engineering Trustworthy Software Systems—3rd International School, SETSS, Tutorial Lectures*, vol. 11174 of *Lecture Notes in Computer Science*, pages 115–142, Springer, Chongqing, China, April 17–22. DOI: 10.1007/978-3-030-02928-9_4 152

Leino, K. R. M. and Müller, P. (2009). A basis for verifying multi-threaded programs. In Castagna, G., Ed., *Programming Languages and Systems, 18th European Symposium on Programming, ESOP, Held as Part of the Joint European Conferences on Theory and Practice of Software, ETAPS, Proceedings*, vol. 5502 of *Lecture Notes in Computer Science*, pages 378–393, Springer, York, UK, March 22–29. DOI: 10.1007/978-3-642-00590-9_27 38, 152

Leino, K. R. M., Müller, P., and Smans, J. (2009). Verification of concurrent programs with chalice. In Aldini, A., Barthe, G., and Gorrieri, R., Eds., *Foundations of Security Analysis and Design V, FOSAD 2007/2008/2009 Tutorial Lectures*, vol. 5705 of *Lecture Notes in Computer Science*, pages 195–222, Springer. DOI: 10.1007/978-3-642-03829-7_7 3

Leino, K. R. M. and Pit-Claudel, C. (2016). Trigger selection strategies to stabilize program verifiers. In Chaudhuri, S. and Farzan, A., Eds., *Computer Aided Verification—28th International Conference, CAV, Proceedings, Part I*, vol. 9779 of *Lecture Notes in Computer Science*, pages 361–381, Springer, Toronto, ON, Canada, July 17–23. DOI: 10.1007/978-3-319-41528-4_20 147

Leroy, X. (2006). Formal certification of a compiler back-end or: Programming a compiler with a proof assistant. In Morrisett, J. G. and Jones, S. L. P., Eds., *Proc. of the 33rd ACM SIGPLAN-SIGACT Symposium on Principles of Programming Languages, POPL*, pages 42–54, Charleston, SC, January 11–13. DOI: 10.1145/1111037.1111042 2

Levandoski, J. J., Lomet, D. B., and Sengupta, S. (2013). The BW-tree: A B-tree for new hardware platforms. In Jensen, C. S., Jermaine, C. M., and Zhou, X., Eds., *29th IEEE International Conference on Data Engineering, ICDE*, pages 302–313, IEEE Computer Society, Brisbane, Australia, April 8–12. DOI: 10.1109/icde.2013.6544834 97, 155

Levandoski, J. J. and Sengupta, S. (2013). The BW-tree: A latch-free B-tree for log-structured flash storage. *IEEE Data Eng. Bull.*, 36(2):56–62. 148

Liang, H. and Feng, X. (2013). Modular verification of linearizability with non-fixed linearization points. In Boehm, H. and Flanagan, C., Eds., *ACM SIGPLAN Conference on Programming Language Design and Implementation, PLDI'13*, pages 459–470, Seattle, WA, June 16–19. DOI: 10.1145/2499370.2462189 108, 151, 153, 155

Manna, Z. and Pnueli, A. (1995). *Temporal Verification of Reactive Systems—Safety*. Springer. DOI: 10.1007/978-1-4612-4222-2 150

Meyer, R. and Wolff, S. (2019). Decoupling lock-free data structures from memory reclamation for static analysis. *Proc. ACM Program. Lang.*, 3(POPL):58:1–58:31. DOI: 10.1145/3290371 153

Meyer, R. and Wolff, S. (2020). Pointer life cycle types for lock-free data structures with memory reclamation. *Proc. ACM Program. Lang.*, 4(POPL):68:1–68:36. DOI: 10.1145/3371136 153

Michael, M. and Scott, M. (1995). Correction of a memory management method for lock-free data structures. *Technical Report TR599*, University of Rochester. 1

Michael, M. M. (2004). Hazard pointers: Safe memory reclamation for lock-free objects. *IEEE Trans. Parall. Distrib. Syst.*, 15(6):491–504. DOI: 10.1109/tpds.2004.8 153

Müller, P. (2001). Modular specification and verification of object-oriented programs. Ph.D. thesis, FernUniversität Hagen. 3, 152

Müller, P., Poetzsch-Heffter, A., and Leavens, G. T. (2006). Modular invariants for layered object structures. *Sci. Comput. Program.*, 62(3):253–286. DOI: 10.1016/j.scico.2006.03.001 152

Müller, P., Schwerhoff, M., and Summers, A. J. (2017). Viper: A verification infrastructure for permission-based reasoning. In Pretschner, A., Peled, D., and Hutzelmann, T., Eds., *Dependable Software Systems Engineering*, vol. 50 of *NATO Science for Peace and Security Series—D: Information and Communication Security*, pages 104–125, IOS Press. DOI: 10.1007/978-3-662-49122-5_2 152

Nanevski, A., Ley-Wild, R., Sergey, I., and Delbianco, G. A. (2014). Communicating state transition systems for fine-grained concurrent resources. In Shao, Z., Ed., *Programming Languages and Systems—23rd European Symposium on Programming, ESOP, Held as Part of the European Joint Conferences on Theory and Practice of Software, ETAPS, Proceedings*, vol. 8410 of *Lecture Notes in Computer Science*, pages 290–310, Springer, Grenoble, France, April 5–13. DOI: 10.1007/978-3-642-54833-8_16 3, 150

O'Hearn, P. W. (2004). Resources, concurrency and local reasoning. In Gardner, P. and Yoshida, N., Eds., *CONCUR—Concurrency Theory, 15th International Conference, Proceedings*, vol. 3170 of *Lecture Notes in Computer Science*, pages 49–67, Springer, London, UK, August 31–September 3. DOI: 10.1007/978-3-540-28644-8_4 150

O'Hearn, P. W. (2007). Resources, concurrency, and local reasoning. *Theor. Comput. Sci.*, 375(1–3):271–307. DOI: 10.1016/j.tcs.2006.12.035 3

O'Hearn, P. W. (2019). Separation logic. *Commun. ACM*, 62(2):86–95. DOI: 10.1145/3211968 150

O'Hearn, P. W., Reynolds, J. C., and Yang, H. (2001). Local reasoning about programs that alter data structures. In Fribourg, L., Ed., *Computer Science Logic, 15th International Workshop, CSL, 10th Annual Conference of the EACSL, Proceedings*, vol. 2142 of *Lecture Notes in Computer Science*, pages 1–19, Springer, Paris, France, September 10–13. DOI: 10.1007/3-540-44802-0_1 3, 4, 150

O'Hearn, P. W., Rinetzky, N., Vechev, M. T., Yahav, E., and Yorsh, G. (2010). Verifying linearizability with hindsight. In Richa, A. W. and Guerraoui, R., Eds., *Proc. of the 29th Annual ACM Symposium on Principles of Distributed Computing, PODC*, pages 85–94, Zurich, Switzerland, July 25–28. DOI: 10.1145/1835698.1835722 153, 155

O'Neil, P. E., Cheng, E., Gawlick, D., and O'Neil, E. J. (1996). The log-structured merge-tree (LSM-tree). *Acta Informatica*, 33(4):351–385. DOI: 10.1007/s002360050048 5, 97, 98

Owicki, S. S. and Gries, D. (1976). Verifying properties of parallel programs: An axiomatic approach. *Commun. ACM*, 19(5):279–285. DOI: 10.1145/360051.360224 3, 37, 150

Parkinson, M. J. and Summers, A. J. (2012). The relationship between separation logic and implicit dynamic frames. *Log. Methods Comput. Sci.*, 8(3). DOI: 10.2168/lmcs-8(3:1)2012 152

Pierce, B. C., de Amorim, A. A., Casinghino, C., Gaboardi, M., Greenberg, M., Hritcu, C., Sjöberg, V., and Yorgey, B. (2020). *Logical Foundations*, vol. 1 of *Software Foundations*. Electronic textbook. Version 5.8, http://softwarefoundations.cis.upenn.edu 150

Piskac, R., Wies, T., and Zufferey, D. (2013). Automating separation logic using SMT. In Sharygina, N. and Veith, H., Eds., *Computer Aided Verification—25th International Conference, CAV, Proceedings*, vol. 8044 of *Lecture Notes in Computer Science*, pages 773–789, Springer, Saint Petersburg, Russia, July 13–19. DOI: 10.1007/978-3-642-39799-8_54 7, 147

Piskac, R., Wies, T., and Zufferey, D. (2014). Grasshopper—complete heap verification with mixed specifications. In Ábrahám, E. and Havelund, K., Eds., *Tools and Algorithms for the Construction and Analysis of Systems—20th International Conference, TACAS, Held as Part of the European Joint Conferences on Theory and Practice of Software, ETAPS, Proceedings*, vol. 8413 of *Lecture Notes in Computer Science*, pages 124–139, Springer, Grenoble, France, April 5–13. DOI: 10.1007/978-3-642-54862-8_9 7, 145, 152

Pnueli, A. (1977). The temporal logic of programs. In *18th Annual Symposium on Foundations of Computer Science, Providence*, pages 46–57, IEEE Computer Society, Rhode Island, October 31–November 1. DOI: 10.1109/sfcs.1977.32 150

Pöttinger, B. (2020). Verification of the flow framework from local reasoning for global graph properties. Masterarbeit, Technische Universität München. 67

Raad, A., Villard, J., and Gardner, P. (2015). Colosl: Concurrent local subjective logic. In Vitek, J., Ed., *Programming Languages and Systems—24th European Symposium on Programming, ESOP, Held as Part of the European Joint Conferences on Theory and Practice of Software, ETAPS, Proceedings*, vol. 9032 of *Lecture Notes in Computer Science*, pages 710–735, Springer, London, UK, April 11–18. DOI: 10.1007/978-3-662-46669-8_29 150

Reynolds, J. C. (2002). Separation logic: A logic for shared mutable data structures. In *17th IEEE Symposium on Logic in Computer Science (LICS), Proceedings*, pages 55–74, IEEE Computer Society, Copenhagen, Denmark, July 22–25. DOI: 10.1109/lics.2002.1029817 3, 4, 150

Sergey, I., Nanevski, A., and Banerjee, A. (2015). Mechanized verification of fine-grained concurrent programs. In Grove, D. and Blackburn, S. M., Eds., *Proc. of the 36th ACM SIGPLAN Conference on Programming Language Design and Implementation*, pages 77–87, Portland, OR, June 15–17. DOI: 10.1145/2737924.2737964 151

Sergey, I., Nanevski, A., Banerjee, A., and Delbianco, G. A. (2016). Hoare-style specifications as correctness conditions for non-linearizable concurrent objects. In Visser, E. and Smaragdakis, Y., Eds., *Proc. of the ACM SIGPLAN International Conference on Object-Oriented Programming, Systems, Languages, and Applications, OOPSLA, part of SPLASH*, pages 92–110, Amsterdam, The Netherlands, October 30–November 4. DOI: 10.1145/3022671.2983999 149, 151

Severance, D. G. and Lohman, G. M. (1976). Differential files: Their application to the maintenance of large databases. *ACM Trans. Database Syst.*, 1(3):256–267. DOI: 10.1145/320473.320484 7, 97

Shasha, D. E. and Goodman, N. (1988). Concurrent search structure algorithms. *ACM Trans. Database Syst.*, 13(1):53–90. DOI: 10.1145/42201.42204 3, 5, 59, 63, 153

Smans, J., Jacobs, B., and Piessens, F. (2009). Implicit dynamic frames: Combining dynamic frames and separation logic. In Drossopoulou, S., Ed., *ECOOP—Object-Oriented Programming, 23rd European Conference, Proceedings*, vol. 5653 of *Lecture Notes in Computer Science*, pages 148–172, Springer, Genoa, Italy, July 6–10. DOI: 10.1007/978-3-642-03013-0_8 152

Svendsen, K. and Birkedal, L. (2014). Impredicative concurrent abstract predicates. In Shao, Z., Ed., *Programming Languages and Systems—23rd European Symposium on Programming,*

ESOP, Held as Part of the European Joint Conferences on Theory and Practice of Software, ETAPS, Proceedings, vol. 8410 of *Lecture Notes in Computer Science*, pages 149–168, Springer, Grenoble, France, April 5–13. DOI: 10.1007/978-3-642-54833-8_9 3

Svendsen, K., Pichon-Pharabod, J., Doko, M., Lahav, O., and Vafeiadis, V. (2018). A separation logic for a promising semantics. In Ahmed, A., Ed., *Programming Languages and Systems— 27th European Symposium on Programming, ESOP, Held as Part of the European Joint Conferences on Theory and Practice of Software, ETAPS, Proceedings*, vol. 10801 of *Lecture Notes in Computer Science*, pages 357–384, Springer, Thessaloniki, Greece, April 14–20. DOI: 10.1007/978-3-319-89884-1_13 151

Turing, A. M. (1949). Checking a large routine. In *Report of a Conference on High Speed Automatic Calculating Machines*, pages 67–69, Univ. Math. Lab., Cambridge. 150

Turon, A. J., Thamsborg, J., Ahmed, A., Birkedal, L., and Dreyer, D. (2013). Logical relations for fine-grained concurrency. In Giacobazzi, R. and Cousot, R., Eds., *The 40th Annual ACM SIGPLAN-SIGACT Symposium on Principles of Programming Languages, POPL'13*, pages 343– 356, Rome, Italy, January 23–25. DOI: 10.1145/2480359.2429111 151

Vafeiadis, V. (2008). Modular fine-grained concurrency verification. Ph.D. thesis, University of Cambridge, UK. 107, 151

Vafeiadis, V. (2009). Shape-value abstraction for verifying linearizability. In Jones, N. D. and Müller-Olm, M., Eds., *Verification, Model Checking, and Abstract Interpretation, 10th International Conference, VMCAI, Proceedings*, vol. 5403 of *Lecture Notes in Computer Science*, pages 335–348, Springer, Savannah, GA, January 18–20. DOI: 10.1007/978-3-540-93900-9_27 153

Vafeiadis, V. and Parkinson, M. J. (2007). A marriage of rely/guarantee and separation logic. In Caires, L. and Vasconcelos, V. T., Eds., *CONCUR—Concurrency Theory, 18th International Conference, CONCUR, Proceedings*, vol. 4703 of *Lecture Notes in Computer Science*, pages 256– 271, Springer, Lisbon, Portugal, September 3–8. DOI: 10.1007/978-3-540-74407-8_18 3, 150

Winskel, G. (1993). *The Formal Semantics of Programming Languages—An Introduction*. Foundation of Computing Series, MIT Press. DOI: 10.7551/mitpress/3054.001.0001 150

Xiong, S., da Rocha Pinto, P., Ntzik, G., and Gardner, P. (2017). Abstract specifications for concurrent maps. In Yang, H., Ed., *Programming Languages and Systems—26th European Symposium on Programming, ESOP, Held as Part of the European Joint Conferences on Theory and Practice of Software, ETAPS, Proceedings*, vol. 10201 of *Lecture Notes in Computer Science*, pages 964–990, Springer, Uppsala, Sweden, April 22–29. DOI: 10.1007/978-3-662-54434-1_36 150

Yang, X., Chen, Y., Eide, E., and Regehr, J. (2011). Finding and understanding bugs in C compilers. In Hall, M. W. and Padua, D. A., Eds., *Proc. of the 32nd ACM SIGPLAN Conference on Programming Language Design and Implementation, PLDI*, pages 283–294, San Jose, CA, June 4–8. DOI: 10.1145/1993498.1993532 2

Zhang, Z., Feng, X., Fu, M., Shao, Z., and Li, Y. (2012). A structural approach to prophecy variables. In Agrawal, M., Cooper, S. B., and Li, A., Eds., *Theory and Applications of Models of Computation—9th Annual Conference, TAMC, Proceedings*, vol. 7287 of *Lecture Notes in Computer Science*, pages 61–71, Springer, Beijing, China, May 16–21. DOI: 10.1007/978-3-642-29952-0_12 107, 151

Zhu, H., Petri, G., and Jagannathan, S. (2015). Poling: SMT aided linearizability proofs. In Kroening, D. and Pasareanu, C. S., Eds., *Computer Aided Verification—27th International Conference, CAV, Proceedings, Part II*, vol. 9207 of *Lecture Notes in Computer Science*, pages 3–19, Springer, San Francisco, CA, July 18–24. DOI: 10.1007/978-3-319-21668-3_1 153, 155

Authors' Biographies

SIDDHARTH KRISHNA

Siddharth Krishna is a post-doctoral researcher at Microsoft Research Cambridge, UK. He completed his Ph.D. in logic and verification at New York University, where he had the good fortune to work with Nisarg, Dennis, and Thomas. He now works on parallel and distributed algorithms for large-scale machine learning workloads. When he is not pleading with computers to do his bidding, he likes to watch comedy panel/news shows and play ultimate frisbee.

NISARG PATEL

Nisarg Patel is a Ph.D. student at New York University's Department of Computer Science, where he works with Siddharth, Dennis, and Thomas on automated verification of concurrent programs. His academic interests also include synthesis of controller programs for robots. Outside of computer science, he loves playing football and reading about history and politics.

DENNIS SHASHA

Dennis Shasha is a Julius Silver Professor of computer science at the Courant Institute of New York University and an Associate Director of NYU Wireless. In addition to his long fascination with concurrent algorithms, he works on meta-algorithms for machine learning to achieve guaranteed correctness rates; with biologists on pattern discovery for network inference; with physicists and financial people on algorithms for time series; on database tuning; and tree and graph matching.

Because he likes to type, he has written six books of puzzles about a mathematical detective named Dr. Ecco, a biography about great computer scientists, and a book about the future of computing. He has also written technical books about database tuning, biological pattern recognition, time series, DNA computing, resampling statistics, and causal inference in molecular networks.

He has written the puzzle column for various publications including *Scientific American*, *Dr. Dobb's Journal*, and currently the *Communications of the ACM*. He is a fellow of the ACM and an INRIA International Chair.

THOMAS WIES

Thomas Wies is an Associate Professor in computer science at the Courant Institute of New York University and a member of the Analysis of Computer Systems Group. His research interests are in programming languages and formal methods with a focus on program analysis and verification, automated deduction, and correctness of concurrent software. He is the recipient of an NSF CAREER Award and has won multiple best paper awards. His fascination with concurrent tree traversals extends to his spare time: he enjoys hikes in the woods.

Printed in the United States
by Baker & Taylor Publisher Services